T4-AIL-449

日本の技術経営

TECHNOLOGY MANAGEMENT IN JAPAN

ABOUT THE PAPERS

This volume contains a set of papers presented at the 1993 AAAS Annual Meeting. It was prepared as a means for recording and rapidly disseminating the information presented and discussed at one symposium. The papers appear as delivered and have not been reviewed by the AAAS. The work described and the opinions expressed in the following pages are those of each author and do not necessarily reflect the official views of the American Association for the Advancement of Science, the Japan Society for Science Policy and Research Management, the institutions with which the authors are affiliated or the sponsoring organizations.

日本の技術経営

TECHNOLOGY MANAGEMENT IN JAPAN

R&D Policy,
Industrial Strategies
and Current Practice

Edited by
Robert S. Cutler

Joint Symposium
AAAS Annual Meeting
American Association for the Advancement of Science
Japan Society for Science Policy and Research Management
Hynes Convention Center, Boston, MA
February 13, 1993

338.952
T255

Contents Copyright 1994 by "Science in Japan" Symposium
All Rights Reserved

Second Printing

Printed in 1994 in the United States of America by Westview Press, Inc.
Boulder, Colorado.

Distributed by American Society for Engineering Education, 1818 N St., N.W.
Suite 600, Washington, D.C. 20036 (USA)   Fax: (202) 265-8504

Library of Congress Cataloging-in-Publication Data

Technology Management in Japan: R&D policy, industrial strategies and
current practice = (Nihon no gigutsu keiei) edited by Robert S. Cutler.
    p. cm. (AAAS "science in Japan" symposium series)
    Parallel title in Japanese characters
"Joint symposium, AAAS annual meeting, American Association for the
Advancement of Science, Japan Society for Science Policy and Research
Management, Hynes Convention Center, Boston, MA, February 13, 1993."
  1. Technology--Japan--Management--Congresses.
  2. Technology and state--Japan
  3. Management--Technology
  I. Cutler, Robert S., 1933-
II. American Association for the Advancement of Science.
III. Kenkyu Gijutsu Keikaku Gakkai.
IV. AAAS National Meeting (1993: Boston, MA)
  V. Title: Technology Management in Japan (Nihon no gijutsu keiei).
VI. Series.
    Printed and bound in the United States of America
    T27.J3T44 1993
    338.95207--dc20   93-8579 CIP
    ISBN 0-9625459-3-7   24.95

For orders and other information, please write to:

SCIENCE IN JAPAN SYMPOSIUM
12306 Captain Smith Court
Potomac, MD 20854 (USA)

Fax: (301) 762-4696

CONTENTS

University Libraries
Carnegie Mellon University
Pittsburgh, PA 15213-3890

Afternoon Session
U.S.-JAPAN INDUSTRY AND TECHNOLOGY MANAGEMENT TRAINING PROGRAM

FOREWORD

A symposium on "Technology Management in Japan" was held in Boston, 13 February 1993, as a part of the 159th Annual Meeting of the American Association for the Advancement of Science (AAAS). It was organized jointly by AAAS Section M (Engineering) and the Japan Society for Science Policy and Research Management.

The theme was selected to improve the general understanding in the United States about the management of technology in Japan. The three organizers—Prof. Fumio Kodama of Saitama University in Japan, Prof. Kazuhiko Kawamura of Vanderbilt University, and I—worked on several projects following my return from a Fulbright Fellowship in Japan. We were convinced that current information regarding Japanese technology management would be valuable to researchers, R&D managers, and technology policy analysts in the United States. Moreover, senior engineering academy leaders in both countries were encouraging and agreed to present the keynote address.

The first section of this volume presents two joint papers which focus on current R&D strategies and industrial practices that differ between the two cultures. A panel of discussants then compares certain Japanese R&D management practices with those now emerging in U.S. industrial operations. The second section records the presentations of four U.S. university programs fostering Japanese language, culture, and management skills designed to prepare U.S. scientists and engineers to work closely with counterparts in Japan. The appendix contains three contributed papers authored by senior R&D directors of leading technology development companies in Japan.

It is my hope that this record documents the ideas expressed by the symposium's speakers and discussants and will be useful to those seeking to gain insights into Japanese technology management practices or to work more effectively with their R&D counterparts in Japan.

Washington, D.C.   Robert S. Cutler
November 1993

ACKNOWLEDGMENTS

This symposium was brought together by the efforts of many people and organizations on both sides of the Pacific, for which I am indeed grateful. Included are: Fumio Kodama and Kazuhiko Kawamura, who served as co-organizers, the ten speakers who prepared the set of papers appearing in this volume, the expert discussants, and the staff of the AAAS Meetings Office.

I gratefully acknowledge H. Guyford Stever, Takashi Mukaibo, Hiroshi Inose, Fujio Niwa, and Richard W. Getzinger for their useful advice and encouragement. I am also indebted to John M. Godfrey of SRI International, who, under support by the AFOSR U.S.-Japan Industry and Technology Management Training Program, prepared the symposium summary and papers for publication. And my special thanks to Sarah T. Cutler, who contributed much at important times.

The session was sponsored by AAAS Section M (Engineering), AAAS Section P (Industrial Science), and the AAAS Directorate for International Programs.

The organizers wish to acknowledge generous financial support from the following co-sponsoring organizations:

Allied-Signal Inc.	Industrial Research Institute
Arthur D. Little, Inc.	Japan Information Access Project
Embassy of Japan	Nippon Steel U.S.A., Inc.
Ford Motor Company	Nissan Research & Development, Inc.
Gillette Research Institute	SRI International
Honeywell Inc.	MIT Int'l Center for Research on MOT
Hitachi America, Ltd.	Vanderbilt US-Japan Center
Intermatrix Inc.	Xerox Corporation
3M	DOC Japan Technology Program

A special appreciation goes to Ambassador Takakazu Kuriyama, U.S. Embassy of Japan, for assistance in making arrangements for the "Nihon Shoku" Japanese-style reception that followed the formal program.

RSC

Dedicated to the next generation of technology managers

INTRODUCTION

Globalization of science and technology has been accompanied by new challenges and opportunities for American scientists and engineers. The climate for cooperation and information exchange has changed, but research cooperation and economic competition are often uncomfortably intertwined.

Although respect for Japanese achievements in technology continues, it has been somewhat difficult to have easy exchanges of information because of language and cultural problems. What has turned the world's attention to Japan over the past three decades, besides its economic strength, is its phenomenal progress in applied scientific research, especially in those fields that are the basis of the engineering of high-tech products and processes now evident in global marketplaces.

It has been my pleasure for the past six years to participate in these joint "Science in Japan" symposium series, every other year. Thanks to efforts of Robert Cutler of AAAS Section M, and to my long-time Japanese colleague Takashi Mukaibo, president of the Academy of Engineering of Japan, I find that these national meetings serve to bring together interested American and Japanese scientists and engineers in the United States. And another thing I have observed is that more and more of the people in the audience—both among the Japanese and the Americans who attend these meetings—are those who in fact actually have interacted in one way or another with counterparts across the Pacific.

This joint Japanese-American conference is not new for AAAS Annual Meetings. The idea was first considered thirty years ago in 1963 when the AAAS Board of Directors decided to hold a national symposium to update American scientists on major areas of research in Japanese laboratories. The purpose was to bring together U.S. and Japanese scientists, stimulate cooperative interchanges, and contribute to international scientific good will.

Several of us on the program today were together for dinner last evening. At one point in the conversation, the topic focused on the latest in electronic communications technology. A curious question was raised:

"Does the extreme of the so-called 'telecommunications revolution' mean that we shall no longer attend such national professional meetings as the AAAS Annual Meeting? And instead of being able to enjoy a meal with colleagues who arrived from distant cities and countries, somehow we would substitute a 'virtual dinner,' produced electronically?"

We then enlarged on the thought of virtual reality as applied to this AAAS professional meeting. First, we agreed that the younger people in our field should not be denied the pleasures of getting from Narita Airport to downtown Tokyo during rush hours or of trying to get into Boston during a typical New England multiple snowstorm. Finally, we developed this collegial theme to the point where one asked, "How could we send Boston scrod by fax to Tokyo, or sushi and sake via e-mail to Boston?" Seriously, I am glad and most relieved that we are physically together today, and that our personal interactions will be actual rather than virtual.

The theme for this symposium—Technology Management in Japan— is particularly fitting at this time, largely because it reflects the successful integration in the recent Japanese experience of two important functions: R&D management and business management.

For years now, many of us have been involved in exchanges between the American, European, and Japanese researchers and policy people. It has been characteristic of those events that, because of our hazy view of the Japanese R&D system, we kind of spiraled in to find a suitable topic for discussion. I think today we finally have arrived at a clear focal point. In the early days, there were mutual questions about basic research and the financing of corporate R&D. That is part of us now. Today, the important questions seem to be about technology management. The facts, of course, are that Japanese companies have established an admirable record in managing high technology for civilian markets and that many Americans would like to learn how.

In the larger sense, I think the theme of this joint symposium is that R&D leaders from two different cultures can learn important things from each other. It's not so much a matter of who is ahead of whom, but rather how to open up the exchange of ideas so that people around the world can benefit from the positive-sum game of having every nation's technology management system improve.

H. Guyford Stever

PROLOGUE

Debates about international trade policy—like the Structural Impediments Initiative (SII) and bilateral trade negotiations—have left Japan and the U.S. feeling ill at ease with each other.

Even though Japanese business leaders are convinced that there are rational explanations for some of the Japanese practices termed exclusionary in the SII talks, they remain frustrated in their attempts to offer explanations that sound rational to Western ears.

On the American side, there are some intellectuals who believe that rationality must lie behind Japan's business practices—if only because they are so successful—but these same intellectuals are disappointed by Japanese explanations. The Americans expect logic, and when their expectations are not met, disappointment turns to distrust.[1]

Discussion in Japan inclines toward the pragmatic. The Japanese often seem to feel that if a system works well overall then—proof positive—it must be a good system. But this approach tends to overlook why things run well, what conditions are conducive to the system's well-being, what is good and what is wasteful—in other words, the stuff of logical analysis.

In the United States, however, discussion cannot even take place without logic. Logic is a prerequisite, a starting point. The interesting part is what follows: dialogue. But to the Japanese, this kind of give-and-take has the appearance of arguing for the sake of argument.

With this basic understanding of the cultural differences, Lewis M. Branscomb and I will attempt to offer a logical explanation behind the technology strategies of certain Japanese companies. We hope that our analytical work provides a starting point for future joint research collaborations between other university researchers in the United States and in Japan.

During the 1991 spring semester, Prof. Branscomb and I decided to start the process. We invited ten senior Japanese R&D executives to lecture at a special seminar at the JFK School of Government, Harvard University. The initial results of the seminar series are presented in this symposium.[2]

Our joint project[3] quite intentionally did not dwell on trade policy or practices; it was restricted to the management of technology within the firm. We hoped that a more rational discussion of management issues would clarify an important part of the analysis of Japanese business practices. There is much interest in the United States and in Europe in those dimensions of Japanese management that were not borrowed from the West, or at least are adaptations unique to Japan.

Japan has had a major impact on the world during this century and should take upon itself the task of completing, by the end of the century, a thoroughgoing analysis of Japanese technology and business practices. In short, Japan owes the world a logical insight into its own achievements.

Similarly, the United States has not only been the engine of both scientific progress and technological development in the period since World War II, but there are signs that U.S. manufacturing has profited from a competitive world trading environment and is regaining market share in some of the most visible high-tech industries. After a period when Japanese management methods were being emulated in the U.S., new approaches representing American adaptations of Japanese practice are now appearing. The concept of "agile manufacturing" put forth by Lehigh University is an example.

We hope this symposium stimulates further study and analyses of the logic of Japanese management methods, and that American firms will attempt to bring objective analysis to their understanding of the roots of success of their trade competitors in Japan, most especially those technology management methods internal to the firm that have proven effective. However trade frictions are resolved, it is the intrinsic productivity growth and innovation rate of the firm that ultimately determine the economic health of a nation.

Perhaps joint researcher teams from other universities in the U.S. and Japan will accept the challenge to continue this important collaborative work.

Fumio Kodama

---

1 Adapted from the author's article "Japan owes the world logic insight," appearing in *The Nikkei Weekly*, Tokyo, Japan (Oct. 19, 1992).

2 Branscomb and Kodama, *Managing Innovation and Setting Technology Strategy in the Japanese Electronics and Energy Industry: Some Insights for Further Exploration*, Center for Science & Technology International Affairs; Science & Technology and Public Policy Program; John F. Kennedy School of Government; Harvard University (January 1993).

3 Branscomb and Kodama, *Japanese Innovation Strategy: Technical Support for Business Visions*, University of America, Inc., Lanham, Maryland (1993).

SUMMARY

Japan's advancements in high technology demonstrate a system of scientific research and technological development that differs in significant ways from that of the United States. Over the past two decades, there has been increased effort by American scientists, industrial managers, and government policy-makers to understand the unique features of Japanese-style R&D. The management of technology has been a crucial success factor for R&D-intensive companies in Japan. Experience in Japanese industry illustrates that successful management of technology requires attention not only to scientific and engineering capabilities, but also to human resources, strategic planning, network building, and the competitive environment. Moreover, technology management requires a focus on innovation from a user's perspective.

As part of a continuing effort to improve the general understanding of R&D practices in Japan, the American Association for the Advancement of Science (AAAS) and the Japan Society for Science Policy and Research Management (*Kenkyu Gijutsu Keikaku Gakkai*) jointly sponsored the symposium "Technology Management in Japan" on 13 February 1993. The program was held as a part of the 159th AAAS Annual Meeting in Boston, Massachusetts. The purposes were to examine key Japanese R&D management strategies and practices and to discuss how American scientists, engineers, and industrial R&D managers might learn the skills needed to work effectively with their high-technology management counterparts in Japan.

The first session of the symposium featured a keynote address, two joint presentations, and the reactions of a panel of three discussants with broad expertise in the study of Japanese technology management. The audience, consisting of some 120 scientists, engineers, and science policy observers from Japan, North America, and Europe, witnessed and participated in a stimulating discussion of the results of recent research on unique features of Japanese technology management. The research focused not only on Japanese high-technology firms taken as a group, but also on the sharp variations in management practices among different types of Japanese high-technology firms.

The keynote speaker was Kaneichiro Imai, vice president of the Japan Society for Engineering Education. Dr. Imai was introduced by H. Guyford Stever, currently a member of the Carnegie Commission on Science, Technology and Government and a former director of both the National Science Foundation and the President's Office of Science and Technology Policy. Dr. Stever also is a member of the Academy of Engineering of Japan.

In his introduction of Dr. Imai, Dr. Stever observed that successive U.S.-Japan symposia at AAAS annual meetings—of which the present symposium was the fourth in a biennial series—have seen growing numbers of people in the audience who have interacted with colleagues across the Pacific in their own academic and professional activities. Dr. Stever noted that Japan's admirable record of achievement in technology management for civilian markets has made Japanese technology management an especially important subject in recent years, but he also stressed, "In the larger sense, the theme is [to find out how] these two great countries…can really benefit from a positive-sum game of having everyone's [R&D] system improve."

Dr. Stever then highlighted Dr. Imai's many accomplishments in a long career as an aeronautical engineer, a professor at Nihon University, a leader in supporting and improving Japanese engineering education, and an active promoter of international cooperation in engineering and research on technology management. He praised Dr. Imai's leadership role in 1976 in winning the coveted Deming Prize for superior quality, and noted Dr. Imai's activities in the Japan Society for Engineering Education, of which he is vice president, the Science Council of Japan, the Japan Society for Science Policy and Research Management, and numerous Japanese professional societies, as well as his involvement as a foreign associate member of the American Societies for Engineering Education, Mechanical Engineering, and Quality Control.

Dr. Imai's presentation, *Technology Management in Japan*, described important changes in R&D management in Japan that have occurred or grown in importance during the past few years—changes that will profoundly influence the character of Japanese technology management into the next century.

Dr. Imai opened his talk by outlining the historical progression of technology development in Japan, moving from "learning" in the post-war period, to "imitation" of foreign technologies in the 1960s and early 1970s, to "innovation" as the Japanese economy matured, and finally to "creation" as Japan developed the capacity to contribute in its own right to meeting the world's need for new technological solutions to a range of problems. Dr. Imai stated that the Japanese government's most recent long-term plan for science and technology policy establishes three goals: coexistence with the earth in an environmentally harmonious relationship, expansion of basic knowledge-oriented intellectual activity, and promotion of a balanced society that is able to meet the needs of the growing number of older citizens. To meet these objectives, Dr. Imai said, the Japanese government intends to double investment in basic research and to increase substantially the number of researchers involved in basic research.

Japan's system of educating its R&D personnel, which includes university-based engineering education as well as extensive in-house training within each company, has proved highly successful in improving Japan's manufacturing productivity in the past. Dr. Imai presented data showing that the decline in students choosing scientific and technical fields in favor of financial and other careers in the late 1980s appears to have reversed. However, promotion of the human-resource base for basic research, as well as a continued focus on engineering and technical training, remains a high priority. Substantial reforms are now in progress in Japanese higher education, including a restructuring of the higher-education curricula on a more decentralized basis than in the past, allowing universities to design their own courses; increases in the number of foreign students at Japanese universities; and an unprecedented, open process of self-review and assessment at the universities. Noting the recent announcement that performance in research will be taken into account in allocation of government funding, Dr. Imai observed, "The market principle [will] be introduced in higher education."

Another priority area in technology management that Dr. Imai outlined was the need for improved R&D productivity, investment, and human resources in small and medium-sized manufacturing companies in Japan. These smaller enterprises—defined as firms with less than 100 million yen ($950,000) capitalization and fewer than 300 employees—account for three-fourths (74 percent) of manufacturing employment in Japan, but only 56 percent of all value added in manufacturing. The Ministry of Labor and the Ministry for International Trade and Industry (MITI) are working hard to improve the development of human resources for R&D in this important sector of the Japanese economy.

Dr. Imai concluded his presentation by outlining the key roles played by globalization of R&D and the need for environmental R&D in guiding technology management in the future. He presented, as a model for responding to these concerns, the five-point declaration of principles for R&D activities by Canon, Inc., which received several major environmental awards in 1992. With globalization and environmental needs serving as the "new vectors" guiding the direction of technology management, Dr. Imai said, "Human resources development—the education and training of the work force for science, technology, and engineering—is the basis of technology management."

Following Dr. Imai's keynote address, Fumio Kodama and Lewis Branscomb made the first of two "cross-cultural" joint presentations, each of which brought together researchers from both sides of the Pacific to examine questions of Japanese-style technology management.

Prof. Kodama is professor at the Graduate School of Policy Science, Saitama University, Japan, and former research director of the Japanese government's National Institute for Science and Technology Policy (Science and Technology Agency). In 1991, he was a visiting professor at the John F. Kennedy School of Government, Harvard University, and in 1992, visiting professor of Mechanical Engineering at Stanford University. Prof. Branscomb is director of the Science, Technology and Public Policy Program at Harvard University's John F. Kennedy School of Government. Formerly, he was vice president and chief scientist of IBM Corporation, the chairman of the National Science Board, and, from 1970-1973, director of the National Bureau of Standards (now, the National Institute of Standards and Technology).

Summary

Kodama spoke first, describing the nature of his collaborative research with Branscomb last year at Harvard on the topic *Technology Strategies of Japanese High-Tech Companies.*

In their research, Kodama and Branscomb developed the hypothesis that, contrary to perceptions commonly held in the United States, corporate approaches to technology management vary more among Japanese industrial firms than among their U.S. counterparts. Kodama suggested that the diversity may be explained by a stronger sense of corporate culture in Japanese firms, due to factors such as the Japanese tendency toward lifetime employment—which lessens movement of personnel among firms—and the relative homogeneity of the MBA management training within U.S. graduate schools of business, in comparison with the substantial amount of in-house training that Japanese managers receive.

Professors Kodama and Branscomb together examined technology management at various Japanese high-technology firms in terms of their particular strategy, structure, and core competence. In his presentation, Kodama suggested a topology categorizing management approaches as market-defined, technology-defined, product-defined, or single-customer-defined. The Sony Corporation, for example, appears to focus its technology R&D on specific products, rather than entire markets. This strategy can be considered as product-defined. By contrast, R&D management strategy at NEC Corporation, a technology-defined firm, seeks first to build core technological competencies within the firm and then to develop products with high market potential that arise from those competencies. The relationship that a single large customer—such as Tokyo Electric Power Co. (TEPCO), the world's largest electric utility—has with its suppliers reflects a supplier-defined technology management strategy; TEPCO'S organization of R&D is characterized by a systems engineering approach, with active participation by its own engineers in its suppliers' R&D programs in order to ease integration of complex systems and to meet its own highly demanding technical and service standards.

Following Kodama's presentation, Branscomb elaborated on several key findings of their research regarding the management of technology in Japanese firms that may surprise many Western observers. Prominent among these is the observation that, although Japanese companies develop technology and pursue new markets aggressively, their technology management practices are more effective at reducing risk than those of many American firms.

Firms defined by core technological competencies, for example, reduce risk by gaining a world-leading understanding of a group of technologies, giving them a sound base from which to pursue aggressive product development and marketing. Another element in assessing and controlling technological risk is the substantial effort devoted in many Japanese firms to ensuring that all managers, including financial and other non-technical officers, have enough understanding of the firm's technologies that they can participate in setting the firm's technology strategy. Prof. Branscomb stated, "I submit to you that the way we [in the United States] manage technology is...a lot riskier than the way the Japanese manage it."

Yoshio Nishi, director of the Hewlett-Packard R&D Center in Palo Alto, California, and Hisashi Kobayashi, the Sherman Fairchild Professor of Electrical Engineering at Princeton University and founder, in 1984, of the IBM Corporation's Japan Science Center in Tokyo, made the second of the two joint presentations in this symposium featuring a combination of U.S. and Japanese perspectives. Both Dr. Nishi and Prof. Kobayashi were educated in Japan and the United States; both participated directly in technology management in both countries; and both have faced the specific task of applying Japanese technology management practices within major American firms in the United States.

Dr. Nishi spoke first on the topic of their joint presentation, *A Comparison Between Japanese and American Technology Management Practices*. His presentation drew primarily on his personal experiences, both currently as R&D director at Hewlett-Packard in California, and previously as research director for Toshiba Corporation in Japan, where he led the widely acclaimed "1-Megabit Dynamic Random-Access Memory" (DRAM) development project. Using the example of Japan's "Very Large Scale Integrated Circuit" (VLSI) design and development program, Nishi illustrated several crucial differences between American and Japanese firms' technology management practices. In contrast to the way U.S. firms operate in a structured "top-down" design approach, starting with computer-aided design (CAD) tools and leading in a deterministic way to a desired design, Japanese firms take a "bottom-up" approach and focus the early planning phase on materials, processes, and manufacturability. Japanese product development cycles involve more concurrent processes, with the result that input from manufacturing, marketing, and even customers can be incorporated readily into the technology management process at a much earlier stage.

Dr. Nishi drew particularly sharp contrasts between U.S. and Japanese technology management with respect to the means of organizing and motivating engineers. A Japanese firm typically places a new hire into a product development team with only a vaguely defined job description; the senior team members play a key role in training and incorporating the new engineer into the team. Performance bonuses are tied to team performance. In addition, he said, Japanese firms seek to motivate engineers by placing a high value on participation in technical meetings, including international conferences, whereas in U.S. firms, participation in conferences serves mainly to aid the individual's personal career development. In many cases, R&D milestones in the product development cycle have been tied explicitly to submission deadlines for technical meetings.

Prof. Kobayashi expanded on Dr. Nishi's observations, discussing the larger role of in-house training, the greater reliance on consensus decision making among R&D teams, and the much greater presence of engineers in top corporate ranks in Japan than in the United States. Referring to Dr. Imai's keynote address, he also noted several disadvantages in the Japanese university-based component of engineering education, such as a rigidly uniform curriculum and the lack of rewards for individual excellence and creativity—a problem in the Japanese corporate R&D setting as well. Prof. Kobayashi posed, as a significant cultural obstacle for Japanese technology managers to overcome, the need to maintain team cohesiveness while also encouraging creativity.

A panel of discussants offered individual reactions to the keynote address and to the two joint presentations. In addition, they made some observations on certain lessons that U.S. technology management scholars and practitioners can learn from their Japanese counterparts.

The panelists were: Martha Caldwell Harris, director of the National Research Council's Office of Japan Affairs; George R. Heaton, a technology policy lawyer teaching at Worcester Polytechnic Institute; and Edward B. Roberts, professor at MIT's Sloan School of Management.

A key point raised by the discussants was the convergence, both in technological strength and in technology management practices, between Japanese firms and those in North America and Europe. Japanese industry is no longer a technology "follower," and many new practices oriented toward basic research, including stronger university-industry ties, are needed. Moreover, the panel also stressed the need for more extensive information and objective data by researchers at U.S. institutions to analyze the different concepts and practices employed in high-technology management by Japanese firms. The particular panel suggestions for further management research include a more complete view of management practices within small and medium-sized Japanese firms, and a study of the interaction between the two different national styles of technology management currently observable within American subsidiaries of Japanese-owned companies operating in the United States.

*　　*　　*

The afternoon session of the symposium featured a panel presentation by directors of four university programs aimed at student training and further study on many of the Japanese technology management methods described in the morning session. It is called the "U.S.–Japan Industry and Technology Management Training Program."

Colleagues at MIT, the University of Michigan, Vanderbilt University, and the University of Wisconsin-Madison have been studying Japanese language, culture, and technology management to learn more from their Japanese R&D counterparts. This activity, supported by the Air Force Office of Scientific Research (AFOSR), is aimed at improving the productivity and global competitiveness of U.S. industry.

The program's objectives are to examine cultural factors observed in Japanese industrial practice, and to transfer technology management methods directly to American scientists, engineers, and managers working in high-tech industries and at certain federal laboratories.

Summary

Claude Cavender, the associate director for Education, Academic and Industry Affairs, AFOSR, began the afternoon session with an overview of the U.S.-Japan Industry and Technology Management Program, which was established in FY 1991. Following this presentation, a panel representing each of the first four university programs to receive AFOSR grant awards, and three industry experts serving as discussants, described and discussed these new university activities in research and education toward the establishment of a new field of academic training called "Technology Management in Japan."

John M. Godfrey

CONTRIBUTORS

Morning Session

KANEICHIRO IMAI, Vice President of the Japan Society for Engineering Education and Professor Emeritus, Nihon University, Tokyo, Japan.

H. GUYFORD STEVER, Commissioner, Carnegie Commission on Science, Technology and Government, and Former Foreign Secretary, National Academy of Engineering, Washington, D.C.

LEWIS M. BRANSCOMB, Professor at the John F. Kennedy School of Government, Harvard University, Cambridge, Massachusetts.

FUMIO KODAMA, Professor at Saitama University, Japan, and visiting professor of mechanical engineering, Stanford University, Palo Alto, California.

YOSHIO NISHI, Director of R&D Center, Hewlett-Packard Company, Palo Alto, California.

HISASHI KOBAYASHI, Sherman Fairchild Professor of Electrical Engineering, Princeton University, Princeton, New Jersey.

MARTHA C. HARRIS, Director of the Office of Japan Affairs, National Research Council, Washington, D.C.

GEORGE R. HEATON, Professor, Worcester Polytechnic Institute, Worcester, Massachusetts.

EDWARD B. ROBERTS, Professor, MIT Sloan School of Management, Cambridge, Massachusetts.

GERALD R. SULLIVAN (Moderator), Consultant, Boulder, Colorado.

TECHNOLOGY MANAGEMENT IN JAPAN

Kaneichiro Imai
Vice President
Japan Society for Engineering Education
Professor Emeritus of Nihon University
Former Member of Science Council of Japan

Ladies and gentlemen and distinguished speakers. It is both a pleasure and an honor for me to be invited to deliver the keynote address of this meeting that has gathered a selected number of the most qualified authorities in the field of technology management. Your presence here in Boston today attests to the importance of the theme of this symposium—"Technology Management in Japan"—and to the eagerness to exchange views to enhance our understanding of it and to further strengthen the climate of collaboration between American and Japanese firms and institutions in this area of vital importance for the economic and social well-being of our respective societies.

Let me try to make the most of this opportunity by being brief and to the point so there will be enough time left for discussion. In presenting the subject, I shall assess current trends affecting R&D policy in Japan, examine several changes in human resources development, and then suggest the "new vectors" for guiding the direction of technology management in Japan.

Japan began the process of restructuring its industries after World War II by importing advanced technology and engineering from more advanced countries. During the almost 50 years since that time, Japanese technological advancement has passed through four stages: learning, imitation, innovation, and creation. Japan has now become a member of the group of advanced-technology-producing countries.

During the first phase, as part of Japan's "income doubling" project, technologies were imported from abroad. Japan selected the best technologies from around the world. Quite simply, the goal was to learn. About ten years ago, when I came to this country to attend a symposium sponsored by MIT, one of the eminent professors at MIT told us that the first step of technology transfer is "learning." Then, in a small voice, he said, "stealing." I was surprised, but I think the feeling of the word may be a little different in English from the Japanese case. My interpretation is that one learns when in school, from professors teaching lessons for which one has paid. But in industry, one tries to "get more than one has paid for"—to maximize the return on learning.

The next steps are imitation and innovation. Innovation, from our perspective, means "step-by-step improvement," not big improvements, just step-by-step improvement. The final stage is creation; in Japan, we believe we are now in this stage. Now we have to work for the transfer of technology outward from Japan.

During the past 10 years, the level of employment in manufacturing in Japan has remained basically the same, while gross domestic product (GDP) has increased. There are many reasons for this phenomenon, but one important reason is that Japan has changed its products to have more value added. For example, in the case of machine tools, Japanese manufacturers have introduced electronic measurement and control. By means of what is called, in Japan, "mechatronics," it has been possible to add more value to the product.

The total quality of manufacturing has improved, for several reasons. The number of graduates from engineering-oriented universities has increased. At the same time, curricula have been changed to meet new requirements of the society as a whole. In-house education in industries has been enhanced as well. The general view of industrial firms is that universities need to provide good students who are well prepared for the future of industrial development. It remains the case, in other words, that industries seek "raw material" from the universities—not "finished parts." This may be different from the attitude in other countries.

Japan's goals in the area of science and technology have evolved continuously over the past 40 to 50 years. On April 24, 1992, the

Government of Japan adopted a new Basic Policy for Science and Technology. This policy embraces three basic principles:

(1) Coexistence of humans in harmony with the earth
(2) Expansion of intellectual stocks
(3) Construction of a "charming" society where people can live with peace of mind.

A correlated priority measure of the Basic Policy, aimed at achieving the basic principles, is to increase the number of researchers in Japan and the overall level of investment in R&D to double the present levels.

Important educational changes have been taking place in university-level education in Japan. The Ministry of Education, Science, and Culture (known as Monbusho) revised the ordinance for university accreditation as of July 1, 1991, in important ways. For the first time, responsibility for creating curricula lies with the universities themselves. In addition, universities must practice self-study and self-assessment, and the results are made available to the public.

Another important need in Japan is for an increase in the R&D activities of small and medium-sized industrial firms. While small and medium-sized enterprises (SMEs), defined as firms with fewer than 300 employees, less than 100 million yen capitalization, account for 99.5 percent of the total *number* of enterprises in Japan and 74.4 percent of the *employees*, they create only 55.5 percent of the *value added*. Both the Ministry of Labor and the Ministry of International Trade and Industry (MITI), the latter through its Agency for Small to Medium Sized Enterprises, are working to increase SMEs' productivity through increased R&D.

A restructuring of the educational system, including courses and curricula, is taking place in Japan. Graduate curricula are improving in both quantity (diversity of offerings) and quality. Refreshment (continuing education) courses for industrial employees are also increasing in number. Japan is making a concerted effort to increase the number of foreign students being accepted each year from around 40,000 to 100,000.

3

These measures represent immediate responses to the challenges facing the educational system in Japan; the outcomes of university self-study and self-assessment activities noted above will produce longer-term results as well. For example, Tokyo University's Science Department has embarked on a review of its activities with the aid of an Assessment Committee consisting of eminent researchers from outside the university. For the first time, government funding allocations will depend in a direct way on the resulting evaluation of the university's performance.

Much attention has been paid in recent years to the shift of young people's interests away from science toward such service-sector fields as finance and insurance. Fortunately, in the past few years, it appears that more students are again embarking on careers related to science, technology, and engineering. Finance and insurance, however, continue to be increasingly popular as well.

In closing, it is important to reemphasize that human resources development, strongly rooted in the education and training of the work force for science, technology, and engineering, is the key to successful technology management. From a long-term perspective, the new dimensions, or vectors, of technology management as seen in Japan are globalization and protection of the environment. The Canon Declaration states these principles clearly:

(1) We reject research development for military purposes.
(2) We do not conduct R&D that is not desirable from an ecological point of view.
(3) We create previously unexplored technologies and product categories.
(4) We respect original technologies and/or product categories created by others.
(5) We conduct R&D activities on a global scale and create new business activities in the country [where they are found].

Thank you.

Kaneichiro Imai

SLIDES IN ORDER OF PRESENTATION

KANEICHIRO IMAI

RATIO of EMPLOYMENT

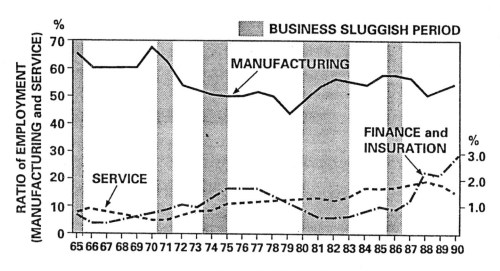

$$\text{RATIO of EMPLOYMENT} = \frac{\text{No. of EMPLOYEE Eng. Course Graduate}}{\text{Total of Eng. Course Graduate}}$$

MANU. POPU. & GDP

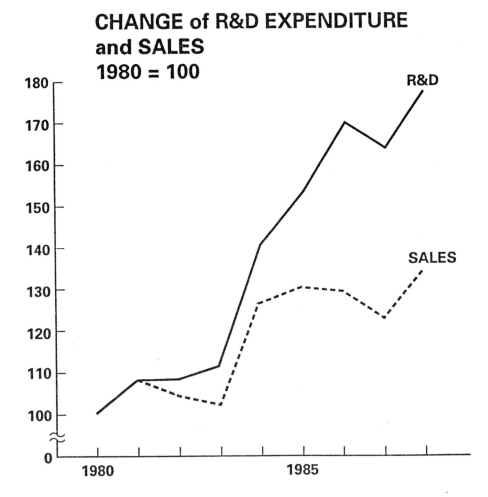

CHANGE of R&D EXPENDITURE and SALES 1980 = 100

(1) We reject research and development themes for military purposes.
(2) We do not conduct R&D that is not desirable from an ecological point of view.
(3) We create previously unexplored technologies and product categories.
(4) We respect original technologies and/or product categories created by others.
(5) We conduct R&D activities on a global scale and create new business activities in the country or territory of its findings.

ENVIRONMENT.   GLOBALIZATION

Progress of Technology Transfer in Japan

Kaneichiro Imai

	Learning 1955	Imination 1965	Innovation 1975	Creation 1985
Social Economic Situation	• Rehabilitation After WW2 • Income Doubling Project	• Open to Int. Market • Environment Problem • Oil Shock	• Increasing International Trade Friction • Steady Economic Growth • Matured Society	• Enhancement Domestic Market • Borderless Economy
Requirement to Science, Technology	• Narrowing the Gap to Advanced Countries • Heavy Chemical Industrialization, and Scale Merits	• Development of Knowledge Oriented Society		• International Contribution
Science & Technology Issues	• Technology Import • Manufacturing/Process Engineering	• Indigenous Technology • Technology for Polution Control • Energy/Natural Resources Tech.		• Enhancement of Creativity • Environment Technology • Tech. Transfer to Overseas

TECHNOLOGY STRATEGIES OF JAPANESE HIGH-TECH COMPANIES[1]

Lewis M. Branscomb
John F. Kennedy School of Government
Harvard University
Cambridge, Massachusetts

Fumio Kodama
Visiting Professor of Mechanical Engineering
Stanford University
Palo Alto, California

Remarks by Fumio Kodama

In this research study, we speculate that there actually may be greater variation in corporate culture among Japanese companies than among their U.S. counterparts. Why might this be the case?

First, one must understand that in the United States employees enjoy a high degree of mobility through their careers, often switching jobs every few years. Consequently American corporate cultures tend to blend into a near-uniform code of behavior and expectations. Moreover, frequent acquisitions and mergers of U.S. businesses and the common training grounds of their executives at the nation's business schools all tend to bring U.S. firms to a common culture.

---

[1] This presentation results from a collaborative project that the two authors have recently completed together. It emerged from a graduate seminar they organized at the John F. Kennedy School of Government, Harvard University, during the 1991-92 academic year. The title of the resulting monograph is "Managing Innovation and Setting Technology Strategy in the Japanese Electronics and Energy Industry: Insights for Further Exploration."

The joint project was made possible by eight Japanese firms: NEC Corporation, Matsushita Corporation, Mitsubishi Electric Corporation, Sharp Corporation, Sony Corporation, Sumitomo Electric Industries, Ltd., Tokyo Electric Power Company, and Toshiba Corporation. Assistance was obtained from senior technical executives at General Motors Corporation and the IBM Corporation, with additional financial support from the Alfred P. Sloan Foundation.

11

What about in Japan? The larger firms there are seen as organic enterprises, with a life and destiny of their own, quite apart from that of their current managers, employees, and owners. Full employment is only one symptom of this culture and of the deep loyalty it commands, reflecting as it does the lower rate of mobility across the corporate boundary. New employees come in not yet fully trained—they don't come into the firm from a business school. The corporation accepts responsibility for their training and indoctrination in the corporate culture. They are indoctrinated in the company, not in the university. This results in the insulation of each firm from the next, allowing for a degree of differentiation that requires great caution about generalizations on Japanese management, technology strategy, and the like. Therefore, one should not be surprised to see substantial differences in the way these Japanese enterprises define themselves.

We have developed a typology for evaluating technology strategy. The typology is based on market, technology, product, and customers. This yields the four categories discussed here: market-defined, technology-defined, product-defined, and single-customer-defined companies.

Although a manufacturing company sets its technology strategy and the organizational management of its technical activities, it may give more weight to one of the categories. Of course, every company must deal with all these factors, but the fact that some give more weight to particular categories leads to strong differences between Japanese companies.

American scholars have done some good academic work describing technology management strategies. I note, however, that these studies are not utilized by U.S. companies, but rather, are more often used by Japanese companies. That may be the case. For example, R. Nelson said that: "corporate strategy is defined as a set of broad commitments made by a firm to define and rationalize its objectives and its intent to pursue them."

Some companies explicitly publish their strategies, but some do not. The important thing to expect is that, once a strategy is defined, the structure of the firm would be adapted to optimize that strategy and the company's core competence becomes adjusted to both strategy and organization.

So here I introduce a typology, which is useful for categorizing the differences we observed in kinds of management and corporate culture. I shall deal with these categories one by one.

The firm defined by its *market* will diversify its technology to guard against unexpected competition unfamiliar to its market. An example is the copper cable manufacturer, Sumitomo Electric Industries (SEI), which sought to protect its markets from the new technology of communications wholly made up of glass fiber.

A firm defined by a particular *technology* may choose to constrain the scope of its technology to ensure a leadership position and will map this advantage onto many different markets. A firm defined by its *product* will stress a specific set of technologies necessary for those products and will specialize in being able to switch from one technology to another. Such a firm is flexible, whereas a technology-defined company is more committed to a particular technology. Finally, a firm defined by *one large customer* will typically develop that technology jointly with its customer.

Market-Defined Company. The market-defined companies develop their business strategy for the demand of customers in globally and functionally defined markets. They will move into any technological area required to increase or preserve their share of their defined market. Sumitomo Electric Industries defines itself as a supplier of communication cable—the same vision it held in 1897, when it started in the copper business. It has diversified considerably over the years, and the company has succeeded in the transition from copper cable to fiber optics.

The model of the technological innovation process at Sumitomo Corporation is referred to as the "bamboo innovation model"—that is Sumitomo's metaphor (see Figure 1). The management principle is based on a 30-year cycle: technology development for the first 10 years, market development for the next 10 years, and then profit making for the last 10 years. If the firm has 100 products, the goal is for some products to be making the profits, which can be invested back into the "technology development" phase. This pattern would be impossible for a company making only a single product.

Technology-Defined Company. The technology-defined company develops its comparative advantage in selected areas of technical know-how, processes, and facilities that largely govern its business strategies. Unlike the market-defined company, the technology-defined company seeks to develop expertise in a few chosen or "core" technologies that are strategic to the business of the company. The strategy is one of concentration, rather than diversification.

NEC is one example. The company's corporate identity slogan is *"C & C Systems: Computers and Communications."* The firm has adopted the metaphor of a tree structure (see Figure 2), with technology at its base. It is interesting to note that, in this model, the "sun" represents the customers. NEC's R&D management is strongly centered around a core technology program. At intervals of 7 to 10 years, NEC assembles a team of research, product development, and marketing people to identify the minimum number of technologies that are necessary and sufficient to capture the maximum number of markets with growing potential. That mind set concentrates on core technologies, rather than the diversification observed in the Sumitomo Electric case.

Product-Defined Company. The product-defined company pursues a discrete set of products whose common characteristics are functional attributes defined from the perspective of the end user. Unlike the market-defined company, whose business objectives will focus on penetrating all segments of a market, the product-defined company will focus on very specific product groups that do not necessarily have a particular, common characteristic. Unlike the technology-defined company, whose products derive from a base of core technologies, the product-defined company creates products that incorporate whatever technologies are required, whether these technologies are available in-house or not.

Sony Corporation is a good example of this type of company. Sony defines itself as "a supplier of audio-visual devices with the finest sound and picture." That is Sony's definition. And Sony assumes its competitor to be the Kodak Company, rather than the other Japanese consumer electronics manufacturers. Sony appears to be a very technologically oriented company, but it does not attempt to cover all the technologies common to the consumer electronics industry. It is significant that Sony is not active in some technologies, such as thin-film transistors, liquid crystal displays, and flat-panel displays, which might be thought of as necessary elements for a technology-driven consumer electronics company. In such instances, Sony's conception of itself led it naturally to stay out of certain technology areas.

Single-Customer-Defined Company. There are companies that deal essentially with a very few customers in a monopolized market, such as electric or gas utilities. Software and hardware suppliers to those types of customers basically deal with a single customer. As a result, they often conduct research jointly with the customer—even very early stages of R&D.

Examples of this type of firm are the suppliers of Tokyo Electric Power Company (TEPCO), the largest private electric utility in the world and one of only nine electric utilities in Japan. Because of its size and position in a highly monopolistic market, TEPCO clearly has a strong hand in the technology provided to it by its suppliers. Moreover, TEPCO itself has 40,000 employees, of whom 25,000 are engineers, including 400 in research. In 1990, its capital investment was $5.5 billion, as compared with the Boston Edison Company with some 4,000 employees and capital investment of about $1 billion. TEPCO also exerts influence by virtue of its extremely challenging technological requirements. One example is the average duration of power outages; in Japan it is seven minutes, compared with outages measured in hours in the United States, the United Kingdom, and France.

Mitsubishi Electric Corporation (MELCO) exemplifies the single-customer-defined company, its major customer being an electric utility. MELCO has been in the business of manufacturing heavy electrical equipment and also supplies the increasingly important operational command and control software for the power distribution networks. This firm has learned that the software development must involve close collaboration between user and manufacturer, as well as the software development team. MELCO developed the "concurrent" model of technological innovation. In this model, users play an important role in all stages, from prototype to marketing. MELCO established an engineering department that develops the architecture of the system and monitors its development, testing, and installation in close coordination with its counterparts in the customer's organization. In this department, 15 engineers play a pivotal role in networking all the relevant players. In the U.S., that might typically be done by an outside consulting firm. But in this case, the engineering function is housed wholly within the manufacturing company.

Remarks by Lewis M. Branscomb

In a 1992 survey, of 95 of the world's leading R&D-intensive firms, Prof. Edward Roberts of MIT's Sloan School of Management, found that senior technical executives in more than 90 percent of the Japanese firms responding had seats on the board of directors, whereas fewer than 25 percent of the chief technical officers in comparable U.S. firms have this level of executive influence. This difference stems from the high level of importance that Japanese corporations place on "technology" and, more importantly, on the "management of technology."

The Japanese firms that Kodama and I studied clearly view "technology" as a major key to corporate success. And they have held that view throughout the postwar period, moving from an era of importing and adapting technology, through competing on manufacturing superiority and incremental improvement, to generating firm-unique technologies via advanced research.

What is striking is that their technology strategies are all well defined and consciously communicated to all their employees through a variety of devices, including such graphical metaphors as the Sumitomo "bamboo" model and the NEC "tree" model. This is true for the market-defined, product-defined, customer-defined, and technology-defined firms, although in each case there clearly are some differences in the technology strategy.

Turning again to the NEC "tree" model, we see the significant point that *technology* is attached to the roots of the tree, rather than to the branches or to the leaves. As Michiyuki Uenohara, former director of NEC's Corporate Research Laboratories, points out in reference to this model, "When the sun is obscured by the clouds" (i.e, the customers are not buying and business is bad), "the roots have to continue to nourish the tree even though the leaves have fallen off." That part of the metaphor says that the firm has to keep driving the technology even through bad times, because that is the core strategy for the company.

The device for new technology development that Uenohara describes is what he has termed the "Diversification and Concentration Strategy." It begins with an assessment from the R&D community of the future potential of all the technologies in their industry, sorting out those that they feel are most promising.   That spectrum of technologies is evaluated by a team that includes the product and marketing people to answer the question of maximizing the number of markets with a minimum number of technologies, for the purpose of focusing the firm's priorities on the core technologies that emerge from that analysis. Then they proceed to develop a set of what are called "model products," involving as much as 20 percent of the central research staff, in order to produce what are in effect manufactured prototypes, not just design prototypes, which are then turned over to the product development section to reconfigure into appropriate products.

The purpose of this strategy of focusing on a defined set of technologies is to reduce the risk of technical failure to a very low level. The firm concentrates its resources on a few technologies, and this emphasis allows it to assemble a world-leading team that should be able to produce a world-leading technical capability. This approach lowers the technology risk, leaving the problem of how to minimize market risk. This is done simply by studying the markets to find the spectrum of markets that is most attractive. This exercise is illustrated in Figure 3, with markets in the rows and technologies in the columns; the goal is to identify that slice through the technology spectrum that best supports the markets.

It may look to the outside observer as though the behavior of this firm is constantly to look at "markets." This is correct, because it already has settled the strategic issue of the "technology" and has moved on to focus on how to address the corresponding full spectrum of "markets."

Uenohara of NEC asserts that "the market alone determines the value of technology and that R&D productivity depends on economies of scope." This is the key to the technology-defined company. He contends also that virtual markets are very important to identify, in a process that Kodama has elsewhere termed "demand articulation." Virtual markets are hypothetical in the sense that, pending realization of the technology, they are not yet articulated, but they arise from perceptions of social and economic need. In that sense, they are more socio-technical rather than simply possibilities suggested by technology alone.

NEC considers R&D's effectiveness to be highly dependent on the way it is managed. Uenohara has said that "technological value is created mostly by management." In this regard, he is including research management, noting that "while corporate management presses for more *efficient* R&D, research management must focus on *effective* R&D."

The above distinction is very important. It differentiates between an efficient R&D management style, which attempts to maximize technical output, and an effective one that places a higher priority on the selection of technical goals and the building of relationships designed to ensure that a successful project will in fact be put to use within the business.

All the firms that we examined in this study (both Japanese and U.S. based) emphasized the management of innovation and its requirement for constant, rapid adaptation. Sumitomo emphasized the importance of an innovation model that incorporates activities before and after the innovation itself, addressing the issues not only of total quality control in production, but also the relationship to the environment into which your market plays.

The firms studied also exhibited a sophisticated view of how technology can be acquired and exploited, and they were not reluctant to share with potential or even actual competitors when the situation warranted. Toshiba Corporation, for example, follows a strategy it calls "C, C and C," for "competition, cooperation and complementarity." Uenohara describes NEC's view as "symbiotic competition," defined as the mutual support and reliance for survival and growth within the basic rule of free competition. During the first 5 to 10 years of a new technology, firms may share with each other through symbiotic collaboration—the stage Western technology policy-makers refer to as "precompetitive research." There then begins a period of free competition, resulting some five years after first product introduction in a new, growing industry.

This concept is relatively unfamiliar to American firms, which have tended to look outside their industry to find the common ground for sharing pre-competitive technology—although it is precisely the policy innovations now going on in the European Framework Program and in many of the R&D consortia that we hear about in the United States that are beginning to get at this question. However, U.S. firms tend to look more to professional and scientific society meetings and trade associations for that mechanism.

An interesting impression that arose from our study is that the aggressive approach to using technology as a strategic asset in business competition for which Japanese companies are famous is in fact buttressed by a conservative approach to the management of technical risk. It was clear that managing technology strategy does enjoy a high priority for top management attention in these firms, higher than in comparable American firms. Many of their strategies seem to be designed to enable non-technical executives, or even executives with engineering backgrounds who have spent most of their careers involved with financial and other non-R&D corporate concerns, to understand the risks and persist, step by step, toward technical objectives with the requisite patience.

We submit the following thesis:

The way American companies manage technology is much riskier than the way Japanese firms manage it.

There is often a tendency among U.S. observers to assume that Japanese managers are less risk averse than their American counterparts, as reflected in the complaints about quarterly profit pressures and risk aversion on the part of CEOs. If the Wall Street financial community would only understand the sense in which investment in technology is a *risk-reducing* strategy for Japanese companies, perhaps they would stop "punishing" U.S. executives for doing the right thing by investing in technology for the long term.

What is the evidence for this assertion? First, the preference of Japanese managers for incremental engineering improvement over the radical-breakthrough alternative as a mechanism for growth is one underlying factor. Through short product cycles and quick response to feedback from the market, both technical and market risk are reduced into smaller and more manageable elements. Second, the attention paid to "reduction to practice" of the prototype products in a realistic manufacturing environment—before they are committed into the business strategy—is another way of reducing technical risk. By contrast, in too many American companies, the engineering design team designs the product, building the breadboard in engineering, and once satisfied that it has the right functional characteristics, "throws it over the wall" to manufacturing and says, "You figure out how to make it."

Third is the continuing search by Japanese companies for signs of emergent technologies arising outside the firm. Market-defined firms like SEI watch for technology arising from industries they are not even in, thus guarding themselves against a strategic assault from outside the industry. Thus, we see the dedication to internal diversification in response to those early warnings as a conservative strategy. Fourth, the "concurrent engineering" model represents a conservative strategy, as well, in that it anticipates process technology problems that arise in manufacturing before the product design phase is complete and the product committed to market, allowing the manufacturer to change the design and maximize "manufacturability."

A particularly interesting item that came up in some earlier work performed at IBM was the Japanese idea of "trickle-up" technology, in which a new technology is first introduced in the company's consumer products, where the volumes are high and the costs are low, before upgrading the function of the technology for use in more expensive and more rewarding applications in business markets.
This practice is a way of focusing first on manufacturing learning, not on optimizing the performance of a new technology.

Finally, another way that Japanese firms reduce risk is by the active pursuit of relevant technical knowledge wherever it may arise. These firms have a much better reputation for willingness to adopt ideas from outside the firm—that is, they are less vulnerable to the "not invented here" syndrome—than their American counterparts. They are paying increasing attention to fostering relationships with technical innovators in the universities—especially U.S. universities. For this, they should be commended; most U.S. universities complain bitterly, not of the Japanese, but of the American companies who are right here, with the technology laid out for them, and don't take advantage of the opportunity—at least, not sufficiently frequently.

It is important to note that the typology of technology management that has been presented herein is a simplification. For example, the Toshiba Corporation, which has been discussed here, was found to reflect in its several businesses every one of the different typology categories. More importantly, we predict that in the future, many of these companies will not remain as narrowly defined as the typology seems to imply. The reason is that all of them are growing, and they are addressing a greater variety of businesses to sustain that growth; and as they expand and form alliances worldwide, they will be required to move toward business strategies that match those of their partners.

Nevertheless, will these trends dilute their distinctive cultural differences, which appear to be unusually well demonstrated in this set of companies? Seiichi Watanabe, R&D director of Sony Corp., has made the case that even as they change the way they do business and use technology, they will retain the corporate uniqueness of their culture and strategy, the individual corporate persona. Watanabe said in a recent letter,

> "I do not think that the globalization of companies will force them to be alike. The necessity of global activities may on the contrary force the companies to establish their own cultures and distinct policies. It can be said the growth and decline of a firm in the long run will be determined by the world's acceptance of the firm's corporate culture and its fundamental policy, which take form as strategies. The emerging global economy will definitely broaden the strategic

options due to the wide variety of customers' tastes and regionality. Thus, the corporate culture and the fundamental policy a firm may select for itself can be chosen from a much broader set of options. In other words, the playing field is becoming greater and wider than ever, creating more interesting scenarios for corporate managers regarding the development of corporate cultures and fundamental policies."

We have undertaken to offer a description, and perhaps even some logical explanations, of the way Japanese electronics, materials, and energy firms manage innovation and set their technology strategies. It is our hope that this description will provide a starting point for a dialogue between Japan and the United States from which each will benefit. The project, quite intentionally, has not dealt with trade policy or practices; it has been restricted to the management of technology within the firm. It is our hope that a more rational discussion of those management issues will help clarify an important part of the analysis of Japanese business practices. There is a lot of interest in the United States and Europe in those dimensions of Japanese management that were not borrowed from the West, or at least are adaptations that are unique to Japan. Japan has had a major impact in the world in this century, and it should take on itself the task of completing, by the end of the century, a thorough-going analysis of Japanese business practices from their own perspective and from the viewpoint of world history. In short, Japan owes the world a logical insight into its own achievements.

Similarly, the United States has not only been the engine of scientific progress and technological development in much of the period following World War II, but there are clear signs that U.S. manufacturing has indeed learned its lesson from a competitive world trading environment and is regaining share in some of the highly visible technology industries. U.S. semiconductor manufacturing, for example, is now growing in market share and has gotten back above the 50 percent mark in world trade. After a period in which Japanese management methods have been studied and emulated in the United States, new approaches that represent American adaptations of the best Japanese methods are appearing. The concept of "agile manufacturing" put forward by the Iacocca Center at Lehigh University is one attractive example.

Our hope is that American firms will also attempt to bring objective analysis to their understanding of the roots of success of their Japanese competitors—especially the management methods internal to the firm that have proved so effective. For, however trading frictions are resolved, it is the intrinsic productivity growth and innovation rate of the firm that will ultimately determine the economic health of both nations.

Lewis M. Branscomb
Fumio Kodama

*   *   *

Question by Stever: "How soon are you going to start analyzing the roots of success of the management methods that have proved so effective in Japan?" Branscomb answers: "The thing to do is perhaps get a team from Saitama University or Tokyo Institute of Technology together like this, *to do that*! Our present work is also subtitled 'Some Insights for Further Exploration.' We hope to carry this project forward as a joint effort with some research collaboration from Japan."

Moderator: "There is also a set of U.S. university groups who are now organizing to do such comparative analyses of U.S. and Japanese technology management policies and practices. We expect to hear from four of them during the afternoon session of this symposium."

SLIDES IN ORDER OF PRESENTATION

FUMIO KODAMA

BAMBOO INNOVATION MODEL

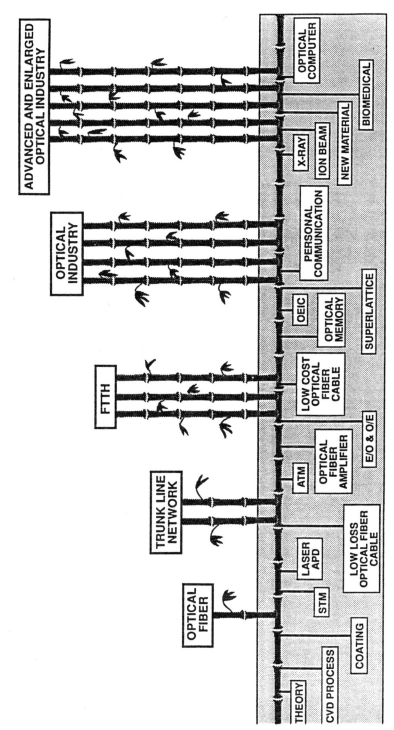

FIGURE 1

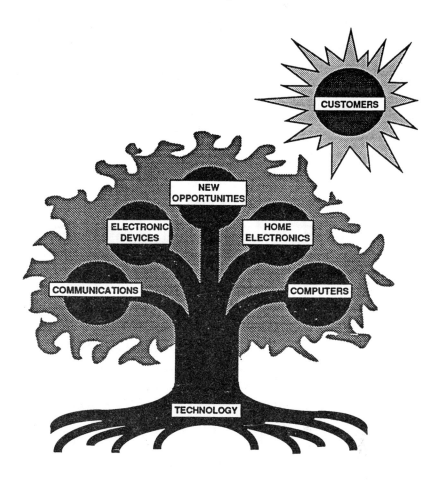

Figure 2 **: NEC Tree**

CORRELATION TABLE BETWEEN CORE TECHNOLOGY AND MARKET

Size / Market	STD	I a	b	c	d	II e	f	g	h	i	III j	··
★ ¥Mill.	1	○	◎	○		○	◎	◎	○		◎	
★	2	○	◎	◎	○	◎	○	○			◎	
★	3	○	◎	◎	◎		◎	○		○	○	
¥	4		◎			○		◎	◎		○	
¥	5	○	◎					○	○	◎	◎	
¥	6	◎	○		○			◎		◎	○	
?	7	○	◎	○	◎	◎	○	○	◎	○	◎	
X	8	○	◎					◎			◎	

FIGURE 3

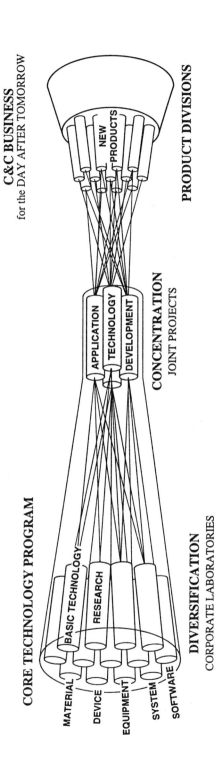

C&C BUSINESS
for the DAY AFTER TOMORROW

PRODUCT DIVISIONS

NEW PRODUCTS

CONCENTRATION
JOINT PROJECTS

APPLICATION
TECHNOLOGY
DEVELOPMENT

CORE TECHNOLOGY PROGRAM

BASIC TECHNOLOGY
RESEARCH

MATERIAL
DEVICE
EQUIPMENT
SYSTEM
SOFTWARE

DIVERSIFICATION
CORPORATE LABORATORIES

DIVERSIFICATION & CONCENTRATION STRATEGY

A COMPARISON BETWEEN JAPANESE AND AMERICAN TECHNOLOGY MANAGEMENT PRACTICES

Yoshio Nishi
Director, R&D Center
Hewlett-Packard Company
Palo Alto, California

Hishashi Kobayashi
Sherman Fairchild Professor
of Electrical Engineering
Princeton University, New Jersey

Remarks by Yoshio Nishi

I am speaking not as a researcher of Japanese technology management, but rather as a practitioner of it. My remarks are drawn from years of experience in the semiconductor industry.

Each generation of semiconductor or integrated circuit (IC) technology typically involves a 6-year lifecycle—the time required to develop and manufacture a new commercial product line, for example, the 1-megabit, 16-megabit or 1-gigabit DRAM devices. However, each new generation of technology should come in approximately every 2.5 to 3 years, according to Gordon Moore's Rule, named after one of the founders of Intel Corporation.

Development Cycle Characteristics

The development cycle for the VLSI and IC devices can be characterized as having two parts: a top-down element and a bottom-up element. The top-down element proceeds from the overall system architecture to a logic diagram to a circuit design, leading to a specific idea of what one wants to design and build. The bottom-up element starts from basic understanding of physics, chemistry, and metallurgy, leading to development of new materials with new properties, development of new processing techniques, and finally to new device structures and/or enhanced manufacturability of devices.

31

At Hewlett-Packard, one observes a much stronger emphasis on the top-down element, which is more a deterministic process and can be thought of as "synthesis by CAD tools." The bottom-up element is more a process of "synthesis by experiment" or trial and error. This is one aspect of electronic product development in which Japanese firms have excelled.

A related distinction can be made between the R&D patterns for application-specific ICs (ASIC) and logic Ics on the one hand, and memory Ics on the other. The former is where U.S. firms such as Hewlett-Packard are doing well; the latter is the Japanese firms' area of excellence. In ASIC/logic R&D, the pattern proceeds from a system concept through logic and circuit design to prototyping and final chip design, and finally to volume production in the sixth year. The process takes place in a "system community" setting, which is a deterministic concept.

The memory IC pattern, by contrast, takes place in a "process community" environment, starting from unit process R&D, followed by determination of the electrical parameters by which one can predict how the prototype chip will perform. Process integration and final design rules are completed by the end of the third year. I should like to point out that by this stage in the development process, patents have been filed and technical reports are presented at international forums such as the International Electronic Devices Meeting (IEDM) and the International Solid State Circuits Conference (ISSCC). Also, customer visits take place at this stage to determine customers' interest in the new product. The final three years of the development cycle encompass process improvement, volume production, and yield improvement, with continued feedback from customers.

In comparison with the former (U.S.-typical) pattern, the Japanese R&D pattern for memory IC chips is more concurrent in nature and achieves demonstration of basic functions in only three years. The subsequent stages focus on improvements to manufacturability. Toshiba's development of the 1-megabit DRAM, which I directed and which was considered almost a legendary victory within the company, gives an example of how the aggressive approach in the first stages of R&D works. The design team started out with a wide range of new technologies to pursue. By years 2 and 3, however, most of these approaches had been trimmed from the design, leaving only two basic innovations, the Planar HiC Cell and Advanced Locos Isolation. This "trimming" process is a key factor, because if the developer attempts to incorporate all the innovations initially considered, the result would be an overly ambitious product. The trimming process leads quickly to a technologically conservative, lower-risk product design that is highly manufacturable.

The question arises whether the careers of the people who worked on those technical approaches that were trimmed from the design suffer as a result. The answer is that these people remain with the firm and continue to develop the technology for the next generation, thus assuring the continuity of flow for the potential technology.

Role of Technical Conferences

Technical conferences play an important role in technology management. For firms in the United States, these conferences are an opportunity for research engineers to demonstrate their technical achievements and their capabilities, share technical information, develop their personal careers, and demonstrate the technical capability of their research organization. Usually, there is not strong support from top management at the CEO or chief operating officer level; and most certainly, the chief financial officer would have "zero interest" in such activity.

By contrast, in Japanese industry, the leading international conferences, especially those held in the United States, have a special significance. Japanese firms see such meetings as an opportunity for free, credible, and efficient advertising, in addition to their generic role as described above. Moreover, they serve as a means for setting R&D milestones on the basis of paper submission deadlines, as well as a venue for sharing new technical information, measuring their own technical progress, and maximizing their engineers' morale and growth. Therefore, these conferences are very strongly supported and encouraged by top management. Participation is used as one of the measures of an R&D manager's capability. At the time of the Toshiba megabit DRAM development, for example, the senior vice president of the company called frequently to ask whether the team would meet conference paper submission deadlines. In fact, the group technical executive in charge of all semiconductor businesses at a Japanese firm usually attends these meetings.

The balance of papers presented at the 1990 IEDM illustrates a contrast between the U.S. and Japanese orientations toward such meetings. In the area of solid-state devices, the 18 U.S.-authored papers split evenly between industry-based and university-based R&D. The 16 Japanese-authored papers all came from industrial R&D. The overall numbers of papers presented by U.S. and Japanese authors were about equal. However, the significance of the U.S. university-based R&D capability in the field is reflected in the strong showing by university researchers at this meeting in the areas of solid-state devices, modeling, and simulation. There were no Japanese papers presented on the basis of research performed at Japanese universities.

University-Industry Interaction

How is university-industry interaction in Japan combined with technology management? In Japan, the universities are viewed as a resource supplier to industry; industry provides funds to universities to support education, but not research in most cases. However, industry does send researchers to serve as lecturers at universities in order to expose students to leading-edge technology.

University research funding comes from the Ministry of Education. It is important to note that the bureaucrats at the Ministry of Education (Monbusho) and at the Ministry for International Trade and Industry (MITI) usually do not collaborate well. Very little technology transfer takes place between universities and industry; however, some recent developments, such as Tohoku University's "ultra-clean technology" laboratory and the Hiroshima University Center for Integrated Systems, indicate change in this regard.

Human Resource Management

Hiring and training of new employees follows a distinctive pattern in Japanese manufacturing firms. Typically, the headquarters unit hires graduates of bachelor's degree programs directly from colleges and universities. These new hires have 1 to 2 weeks of introductory training followed by 2 to 6 months of training at manufacturing and sales sites. Because of this process, even a graduate from an "ivory tower" institution soon understands that his or her research is supported, ultimately, by the company's manufacturing and sales activities. Finally, the employee is distributed to a job site assignment. The supervisor at the R&D lab, in short, does not recruit his new employees directly.

Salary and bonuses are distinctive in Japanese firms as well. Salary is based on individual capability, future potential, and achievement in a manner consistent with promotion ranking. Bonuses are based on achievement and whether an employee shows conscientious effort; the bonus is strongly skewed to team achievement. Good team results lead to promotion of the team leader to the next management level, thus making room for another employee to be promoted as leader. Members of successful teams can enjoy higher bonuses than the company average.

The methods of hiring, training, and rewarding employees are related to some clear distinctions in the "mentality" (attitude and way of thinking and behaving) between R&D engineers in the United States and in Japan. Broadly speaking, U.S. engineers seek a clear, detailed job description and then say, "This is my job; that is your job." The Japanese counterparts have much vaguer job descriptions and might typically say, "This is *our* job." While rewards in the U.S. accrue to the strongest individual player, by contrast in Japan, the rewards go to the whole team.

The vagueness of job descriptions for the new employees in Japanese R&D units (which could lead to a fiasco for U.S. employees) produces interesting and important effects in the Japanese case. The new Japanese employee is forced to talk to his or her colleagues and ask many questions. The group, because it is evaluated on the basis of team achievement, has an incentive to guide and train the new employee effectively. Because technology development is a dynamic process, plans change rapidly. That situation can leave areas uncovered in an organization with rigid job assignments. Moreover, the Japanese model produces a flexible group organizational structure in which members cover and support each other with positive criticism.

The 6-year VLSI product development pattern I described earlier (technology generations come every 2.5 years) involves a great deal of overlap between the research, design, and manufacturing phases. Typically, there is substantial mobility of engineers between research, development, and manufacturing. Approximately 10 percent of the research engineers move to the development area each year; 10 percent move from development to manufacturing. Although there is risk and expense associated with overlapping elements of R&D, following this pattern has made it much easier over time to build up knowledge and "trust" in the firm's view of the prevailing technology trend.

The career path of the typical R&D executive in Japan underscores many of the unique features of the Japanese firms' approach to R&D. After joining a firm at age 24, the employee proceeds through the ranks of management while developing expertise—and finally becoming one of the key experts—in several technology areas. Even at the level of laboratory director or chief engineer, the typical executive remains highly active in technical conferences, technical publishing, and teaching at universities. This level is only one step removed from chief executive positions, illustrating why Japanese companies have so many more technical people in top management compared with U.S. counterparts.

In summary, perhaps we can characterize the differences in managing R&D organizations in the United States and in Japan in terms of entropy. The U.S. model involves less interaction at the individual level, greater individual freedom, more physical separation, and more age equality—resulting in a "hetero-culture" with greater "system entropy." In contrast, the Japanese model exhibits stronger individual interactions in the schools and in the firm, less emphasis on individual uniqueness, and more spatial closeness, and incorporates an age-status hierarchy that values the accumulation of experience—resulting in a "uni-culture" with less entropy.

Therefore, from the technology management point of view, I often struggled to *suppress* the ever-increasing entropy of my organization at Hewlett-Packard; while before, when I was in Japan managing a corresponding research organization at Toshiba, I tried to *increase* the system entropy.

In this light, you may well ask whether it is possible to develop a technology management capability through education and training—one that is objectively measurable and independent of age and experience. I would answer "yes" for the United States, because what we need is to help and coordinate the organization to contain the system entropy. However, for managing an R&D structure in Japan that exhibits much lower system entropy, a technical manager is needed not so much to coordinate and control, but rather to encourage the unit to achieve the new and the original. This is why management training in terms of coordinating and organizing is not as appropriate for Japan, and the relevant management capabilities probably are *not* trainable.

Just as U.S. companies need to strengthen their capabilities for organizing and coordinating R&D activity, Japanese firms need to learn how to try radically new and completely different approaches. It is an important value of this symposium that we could identify and compare the strengths and weaknesses of each side, rather than simply accepting the notion that We should learn from Japan.

Thank you

Yoshio Nishi

SLIDES IN ORDER OF PRESENTATION

YOSHIO NISHI

Period for one generation of technology:

6 years

New technology generation comes in every

2.5 – 3 years

HORIZONTAL TECHNOLOGY INTERACTION

	U.S.	Interaction Frequency	JAPAN	Interaction Frequency
Technical Meeting	MTS	2/year	MTS	1–2/year
Technical Program Committee	Senior MTS	2/year	Section/Department Manager	3–4/year
Steering Committee	Senior MTS Manager	2/year	Department/Lab Manager	5–6/year
Executive Committee	Senior MTS Manager	2/year	General Manager for R&D	5–6/year

Models for Equipment R&D and Commercialization

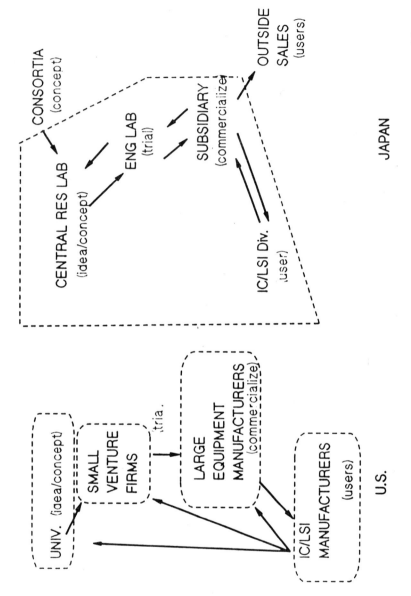

Mentality of Engineers at R&D

U.S.

clear job description

"this is my job and that is your job"

reward to the best player

"strong management skill is required in order to anticipate and refill interdisciplinary gap, which makes manager too busy with daily work"

JAPAN

vague job description

"this is our job"

reward to the team

"less requirement for management skill"
"manager can spend more time for forward looking, longer range issues & strategy"

YNISHI2

43

Vague job description for a new employee:

→ need to talk and ask what/how/where/when to do

→ group, since it will be viewed by team achievement, has incentive to train/guide/help new employee become a contributor

→ flexible organizational structure where members cover/support each other with positive criticism

yn:jobdes

TYPICAL R&D EXECUTIVE IN JAPAN

Join a company as an engineer/scientist — at age 24

First line level manager after establishing
(supervisor)
himself as a technical expert in 1–2 related areas — at age 31

Second line level manager after broadening his
(section manager)
technical areas covering 2–3 related areas — at age 36

Third line level manager after establishing
(department manager)
reputation as one of the key technologists
in several areas — at age 41

Lab Director/Chief Engineer still very active to
participate in technical conferences, teaching
at universities — at age 46

After all, it is common to have 30–60 patents, 30–70 papers and
several books

YNISHI1

Typical Technology Transfer Pattern in Japan

- Define research objectives jointly with D & M

- Transfer 1 ~ 2 engineers from D to R
  Work for 1-2 years

- Return D engineers to D, and transfer 1-2 engineers from R to D

- Transfer 2-4 engineers from M to D
  Work for 1-2 years

- Return M engineers to M, and transfer 2-3 engineers from D to M

Hiring and Training of New Employees

College graduates (BS, MS, PhD)

—Corporate headquarters hires directly from colleges
1–2 weeks introductory training

- corporate policy/hoshin/performance
- business basics
- tour of all plants and labs

2–6 months

- training at manufacturing sites
- training at sales sites

distributed into final sites

job assignment

ynthiring

47

Horizontal Communication and Trust

* *Centralized hiring from colleges, high schools*

* *Training at a central location for ~6 months*

* *Every new hire knows each other after being sent to different organizations at different sites*

* *Organizational collaboration*

**Circuit Technology
Group**

:hcatgal

**HEWLETT
PACKARD**

Remarks by Hisashi Kobayashi[1]

When comparing the technology management systems of the United States and Japan, one key aspect is human resource development. This includes both university education and employee training within the corporation. I do not wish to dwell on the problems we have in higher education in the United States. We currently have a system that is the envy of the world, although a recent study by the President's Council of Advisors for Science and Technology (PCAST) reports that U.S. universities are now facing some disturbing problems. And pre-college education has some well-known problems as well.

However, Japan's educational problems are different. First, the democratic and egalitarian values that have been so firmly established in the culture since World War II have resulted in a strong resistance toward distinguishing individuals in terms of personal capability. Curriculum is rigid, not flexible, leaving no room to encourage gifted students. And at the same time, tremendous pressure is put on less-talented students to keep up with the rest of the student body.

There are some positive aspects of the Japanese educational system as well, such as the high expectations on the part of teachers and parents for every student to perform well. Nevertheless, the egalitarian cultural value is significant in the Japanese school system. Students receive grades, but there is no formal award system for students who excel. There is no "honors" recognition. The Japanese educational system works to develop cohesive team effort in the society at large, but at the same time it suppresses individual creativity. I think this presents a dilemma that Japan must learn to cope with: how to maintain its highly productive, team-oriented work ethic, which is firmly established, and at the same time encourage the interests and talents of individuals to enhance creativity. This remains a difficult and challenging situation to be resolved.

---

[1] The speaker worked 15 years at IBM Research Center in Yorktown Heights, New York, during the period when Dr. Lewis M. Branscomb was vice president and chief scientist of the IBM Corporation. In 1982 he was assigned to IBM Japan to establish the IBM Japan Science Institute in Tokyo, now the IBM Tokyo Research Laboratory, and served as its founding director.

Education in the Japanese firm is based on very extensive and comprehensive company orientation programs. The length varies with the individual company. Typically, it takes from six months to a year to educate new employees coming out of the universities. In my own experience as an employee at Toshiba, for two years between my master's degree and beginning doctoral studies, even master's degree employees were required to spend two months working in manufacturing plants—in the machine tool shop, for example—at various locations. Another example is at Matsushita, where new employees are typically sent to work in ordinary retail electronic appliance stores to sell the company's products directly to customers. This is all part of the very comprehensive company training for Japanese employees.

Recruitment into Japanese firms is also quite different from U.S. counterparts. New employees are initially selected by the corporate headquarters. Then, depending on the ability and interests of individuals and the needs of different divisions, the central personnel department assigns the new employees to the various divisions. By contrast, at U.S. firms, recruitment is usually carried out in "bottom-up" fashion, identifying a candidate locally, who then joins the company with final approval of management and the personnel office at corporate headquarters.

Assignment of research employees and young engineers to universities for graduate study or technical training is yet another key feature of Japanese corporate practice. Assignments at both U.S. and Japanese universities are common. I think that perhaps more U.S. corporations ought to use such academic assignment programs to provide their employees with necessary retraining and other learning mechanisms, which are also effective for conducting technology transfer.

Apart from the human resource issue, Japanese corporate R&D management and strategy are rather distinctive as well. Dr. Branscomb and Prof. Kodama this morning highlighted the fact that Japanese companies spend significant amounts on R&D with very long-term commitments. This is exactly the *reverse* of the situation some 26 years ago, when I first joined IBM. At that time, I was constantly visited by Japanese survey teams and corporate executives seeking to learn how IBM managed its R&D. They were impressed with the ways IBM and

AT&T Bell Labs conducted fundamental research, whereas their Japanese companies engaged only in very short-term product development.

Today we hear exactly the opposite story. The truth may be somewhere between the two extremes. As Dr. Branscomb pointed out earlier, the Japanese technology management style is conservative and tries to avoid risks. And many U.S. companies still do important fundamental research. Nevertheless, the trend is now shifting in the direction of greater Japanese corporate commitment to longer-term basic research, whereas U.S. companies increasingly are driven by short-term profit or revenue targets.

Although Japanese society is still hierarchical and seniority oriented, management style is not really "top-down." People who are old are respected, but they do not really make big decisions by themselves. In some sense, they delegate the key technical decisions to middle managers. Of course, the final decision or announcement might be made by top management, but the real decision-making is usually done by technical managers.

The influential role of the middle managers in Japanese corporations is very important to understand. The technical plans and decisions are made by those who are very active in technology development and very informed about world R&D developments, and who attend many technical conferences. Thus, top executives seem to rely heavily on input from the middle managers.

Certainly, any organization has difficulty achieving cooperation among different groups. And as Branscomb pointed out, Japanese companies are less vulnerable than American firms to the so-called "not invented here" syndrome, which rejects new ideas originating from outside sources. This attitude serves to make technology and new product development efforts more consensus oriented, and characterized by its long-term continuity.

The higher number of engineers and scientists among Japanese electronic company CEOs is yet another difference in the style of management in Japan and the United States, where most companies are run by people with MBA, finance, or marketing training.

In terms of corporate management style in general, as already noted, Japanese society is very hierarchical and seniority oriented, but it is not "top-down." This is also true for government organizations. When a new Prime Minister appoints new cabinet ministers, and so forth, government policies do not really change. This is because the top person, who will be in that position for only a short term, does not make that many influential decisions. Instead, he or she depends on recommendations of the career bureaucrats who have been running the ministries for many years. There is very little discontinuity. By contrast, in the U.S., within the government, the universities, and private industry, there is a constant desire for change. American culture, for better or worse, seeks a large "entropy," as Dr. Nishi observed. And when a new person enters the organization at the top, there is expectation of change.

The desire for new leadership is not as explicit in Japan. The Japanese people by culture and tradition tend toward conservatism and resistance to changes. This can have undesirable aspects, but in terms of continuity and commitment to research and technology development, it is quite positive. The person at the top might wish to make drastic changes, but such ideas would be met with strong resistance.

Finally, an observation about corporate salary structures in the United States and in Japan. Long before it was recently reported in the mass media, I was surprised to see the tremendous disparity in salaries between the top and bottom levels in U.S. firms. I believe that this is related to the notion of accepting the leadership of "stars" or very talented individuals; whereas in Japan, there is the practice of collective decision-making and consensus "leadership from the bottom up." In Japan, a secretary with a bachelor's degree will receive a salary similar to that of an engineer with the same degree. In other words, it is not job description that determines salary, but rather age and seniority.

(Educational attainment matters, too; however, a Ph.D. graduate newly entering a company might receive sometimes a lower starting salary than the salary being paid to someone with less education who has been with the company four or five years.) Overall, the disparity in compensation within the Japanese organization is much smaller than in the United States.

In summary, these remarks describe some of the qualitative differences between Japanese and American technology management practices that I have observed during my career in both cultures. It is difficult to say which system works better. But certainly the strong spirit of teamwork and employee dedication within Japanese firms, and the recognition by top management of the need to work continuously at technology development and technology transfer, are two notable differences.

Thank you.

Hisashi Kobayashi

DISCUSSANT PANEL

Gerald Sullivan, the moderator, opened the discussant panel by thanking the presenters for good insights and authoritative descriptions of Japanese and American approaches to technology management. He emphasized the importance of making such comparisons. First of all, comparison provides a benchmark for knowing what we are working from. Then it raises the question of whether our technology management performance is not the same as our counterparts in Japan. And finally, after asking those questions, perhaps we may be open to learning.

He cited two purposes to this symposium. First, a newly emerging field called "Management of Technology" is beginning to develop in university settings. It combines several important aspects of two traditional academic disciplines—engineering management and business administration—and is fast developing into a recognized field of industrial practice. This symposium can help its development.

The second purpose is to acknowledge the achievement of our Japanese colleagues in managing the rapid development new technologies for civilian purposes. Thus, the symposium can help to wake up American educational and business institutions to the fact that Japanese companies manage technological innovation differently from the way it is done in the United States.

Perhaps we might learn some useful ideas from them.

DISCUSSANT

Martha Caldwell Harris
Director, Office of Japan Affairs
National Research Council
Washington, DC.

I want to discuss three topics and refer back to some of what we heard this morning. The overarching question is: What can we learn from Japan's success in managing technology? One way to answer this question is to observe that the United States and Japan are on two separate trajectories in terms of management of technology: Japan is moving toward a stage of more innovation and creativity, as Dr. Imai explained; the United States is now grappling with management of technology in order to support civilian industry.

The optimistic scenario is that these two trajectories will intersect in mutual learning, as Japan becomes more oriented to outward technology transfer and creative innovation, and as the United States lays more stress on technology acquisition and diffusion. The heart of Japan's corporate experience with management technology, summarized by Dr. Imai, is "systemic innovation" or learning to learn continuously. Although Japanese approaches must be carefully selected and adapted to suit the U.S. context, the U.S. companies can learn from Japan about system innovation.

But I would emphasize that there are many factors that make it unlikely that the two systems (U.S. and Japanese) will "converge." Dr. Hajime Karatsu has emphasized the importance of "keeping technology in your own hands." Encouraging technology transfer, particularly arms–length licensing, by Japanese companies to U.S. firms is not a simple process. In addition, U.S. organizations will need to develop more systematic approaches to accessing and using technology developed abroad. Deep convergence is also unlikely because of the complexity of technology management in Japan and the fact that we do not have a full understanding of the process. Finally, once we get the facts straight about how Japanese companies manage technology, there is another critical step needed to translate the findings into meaningful conclusions and policy recommendations.

Now I would like to make a few comments on some of the difficulties in studying Japanese management of technology. Despite the reams of reports on Japan, our database is still limited. We tend to study a few prominent examples or "success stories" and assume that they represent Japanese approaches more generally. The work of Drs. Branscomb and Kodama makes a contribution in this regard in highlighting a variety of Japanese corporate strategies or success stories. We need to study the failures as well. As Mark Fruin elaborates in his fine book *The Japanese Enterprise System,* the Japanese are not a tribe and Japanese firms not a patch. We need to constantly ask ourselves how representative is the case and appreciate the complexity. For example, although some Japanese managers may be more risk averse than their U.S. counterparts, there are exceptions such as those who in the late 1980's moved to diversify their business activities with large, new investments in areas like biotechnology. Some of those investments were in the United States and did not pay off. What makes some firms willing to take such risks?

Another question we have to deal with as researchers relates to access. How much capability does the researcher have to select source material, and how much direct access does he or she have to the actual workings of Japanese laboratories and production facilities? Of course, our ability to gain significant access depends on our Japanese language capabilities.

Still another problem is the timeliness of the information we gather. Today we hear much about the Japanese "bubble" bursting. Certainly, there are signs of change. But there are also continuities in that Japanese firms see adaptation as a basic fact of life. We need to study change and adaptation in the context of continuity.

Finally, we must acknowledge that there is no accepted methodology for studying Japanese management of technology. Instead, we see a dynamic field that draws from many disciplines. There are, as you know, striking differences in the perspectives of those from the technical and business management streams in the United States. In Japan, there are clearly many who know how to manage technology well, but few who study this process in a systematic way.

So why should we study Japanese management technology, given all these difficulties? This brings me to my third theme: the interface between the intellectual enterprise and the need to inform industry and national policy at a macro level. Looking at the question from this perspective, there are some issues that probably deserve more attention in our study of Japanese management of technology. One is the role of smaller firms. We heard Dr. Imai talk this morning about the productivity of smaller firms in Japan and how this has evolved over time. There is a fascination today in the United States with the *Kosetushi* and mechanisms for technology diffusion that could be informed by a careful study of Japanese practices. Another related topic that deserves attention is interfirm alliances or corporate networks. These linkages are the glue for cooperation in the midst of competition in Japan. These networks in Japan allow for adaptation by combining subdivision of tasks and a refined distribution of rewards.

Finally, we need to look more carefully at patterns of government–business interaction in policy–making. Examples such as the JRDC (Japan Research and Development Organization) illustrate how the Japanese government has for many years provided incentives for commercialization of technology by industry. As Dr. Imai pointed out, Japan's capacity to cope with structural change relates to the crucial role of the government (and MITI) in macro–micro feedback loops.

DISCUSSANT

George R. Heaton
Professor of Management
Worcester Polytechnic Institute
Worcester, Massachusetts.

I am a lawyer by training. Most of my career has been working with, talking to, and teaching engineers and managers—now at Worcester Polytechnic, and previously at MIT. My background focuses me and therefore my remarks today on institutions, public policy, and processes.

A series of very fortunate coincidences have made me into a quasi "insider" in certain aspects of Japanese life. This includes the Japanese family, a Japanese company that I've been part of, a year in a Japanese university with Prof. Kodama at Saitama University and, to some extent, inside a Japanese government agency—the first foreign scholar at the Ministry of Health and Welfare. And most recently, as a participant in a new research project going on in Japan under the JCIP (Japan Commission on Industrial Performance). This project is taking a critical look at the institutions, public policies, and the technology management process in Japan.

I think that many of the things we have heard described today—the system of Japanese technology management and their public policy institutions—have been based on an assumption about the relationship between Japan and the outside world where Japan essentially is a "follower." These institutions have been based on a sense of stability for the relationship— that it would persist over time. Because these conditions now have fundamentally changed, I think that much of the past paradigm upon which Japanese technology management and public policy was based in technical areas, has played itself out. We are at an historic watershed. Japan must do something different as it cannot succeed in the way that it has in the past.

A few words about the old technology management paradigm in Japan and the situation which I believe is now changing it. In describing

61

this paradigm, I want to give credit to three colleagues in the previously mentioned JCIP project—to Prof. Fumio Kodama and William D. Commings, of Harvard, and to Tatsu Suzuki at MIT. We see the old paradigm, particularly of public policy in Japan, as having four parts:

First, there was a high degree of centralization in policy decisions about technology. We have seen this, for example, in the extreme importance of the University of Tokyo, in the central importance of the city of Tokyo, and in the centralized decisions about allocations of capital, industries to be targeted, and the licensing of different types of technologies permitted to come into Japan.

The second part of the paradigm is the emphasizes the supply of technology, the acquisition of technology from abroad, and production of a supply of technically trained people from the universities.

The third aspect of the paradigm is a reliance on the linear model of technology development, which proceeds from basic science to commercialization and from which the Japanese are able to extract particular value in the last stages of that model. The final aspect is what I would call "intra-linkages;" that is, emphasis on intragroup communication. We have heard today of the personnel rotational system and the very good kinds of communication understanding of an institution which the people in it have. And yet we have also seen how hard it is for one institution to relate to another.

Now, why must these things change, if they have worked so well in the past? One reason, I think, is that the Japanese in many areas of technology are in essential parity with the rest of the world. The situation has changed, so the fundamental assumption on which the system was based with respect to the external world must also be changed.

Second, technology development itself may be quite different from the past. Prof. Kodama's notion of "technology fusion," is a correct representation of much that is now happening in the world. It differs fundamentally from the "linear model" of technology development.

Third, we have heard already about globalization and the need for institutions from any country to engage those from all other countries to a greater extent. We see as well that the other nations simply will not stand for the kind of relationship which Japan has assumed with the external world in the past. I think this is very clear in the United States. It is also becoming apparent in places like China, Korea, and Southeast Asia, where there is great demand for changes in their historical relationships.

Last, I think it is very interesting to see the Japanese people themselves and their patterns of social interaction changing dramatically. You can see this in male–female relationships, particularly among the younger part of the population. You can see it in terms of attitude toward authority between parents and children and among people in institutions. However, I am unsure about how Japan in going to change. We have learned that there is a great diversity within Japanese firms—a kind of great intellectual awakening of self criticism, in Japan today. Many proposals are being advanced, some are reformist and some highly radical.

In summary, let me repeat my basic point, which is the historic watershed Japan is now facing. Never before has Japan seen the need to lead the world in the development of technology. This represents an important experiment for the Japanese people.

DISCUSSANT

Edward B. Roberts
David Sarnoff Professor of Management of Technology
MIT Sloan School of Management
Cambridge, Massachusetts.

Good morning. I want to make some brief remarks about technology management in Japan which are based upon a new study, jointly carried out by the MIT Industrial Liaison Program and the PA Consulting Group. My role was to be the bridge between the two organizations. I chair the MIT activities in Management of Technology and Innovation, and also chair the Pugh Roberts Associates which I founded some years ago. Lewis Branscomb mentioned some of the findings of our study in his presentation earlier this morning.

The study was carried out by surveying the 244 largest R&D spenders in the world, those companies in the United States, Japan, and Western Europe whose total R&D expenditures cumulate to more than 80 percent of the total spent on research and development in these major competitor–collaborative market structures. Better that 40 percent of the companies surveyed responded. It is important to note that there was no participation by any small or medium–sized companies; therefore, the bias is to perspectives of very large companies. The minimum R&D spending of any of the firms included in the study was more than $100 million per year. We are talking about large expenditure activities. I emphasize that these findings are preliminary.

The first point amounts to one thing: that at the top level—the strategic management level—Japanese firms appreciate the strategic aspects of technology management and they tie their technology strategy more closely to overall corporate strategy and corporate objectives than large U.S. firms do. Moreover, in my judgment, Japanese firms do a better job of coping with new technology and integrating it into the rest of the corporation.

This morning, Kodama and Branscomb showed us what those strategies were and the range of their diversity. One thing they emphasized was the clear communication to the outside world—and to the employees of the firms—that such strategies existed and <u>what</u> those strategies were. I would challenge corporations in America to demonstrate that such level of clarity of communication exists, because I do not believe that their corporate strategies are well understood among them. One cannot communicate clearly what one does not understand.

The second thing that is critical is the role of the chief executive officer (CEO). We looked at the participation of the CEO in four different dimensions of technological content of strategy (development of strategy, selection of major projects, resource allocation of internal technology, selection of external technology investments). In Japan, we found a dramatically different participation rate across the board than we did in the U.S. Thus, it appears that Japanese CEOs are comfortable with high levels of participation in the content elements of technology. By contrast, American CEOs surpass their counterparts throughout the world with respect to their participation in setting R&D budgets. The American CEO is that of the "cost controller" for R&D—to make sure the budgets are tight.

Third, Lewis Branscomb mentioned this morning that our data demonstrate a dramatic difference in membership of the managing board, vis-a-vis chief technology officers, in Japanese, European, and American companies. In fact, over 90 percent of Japanese firms are represented in their top management councils by the senior technical officer of the firm, whereas half that number are evident in Europe and only a quarter in the United States. The point is that the perspective of the most informed and most senior spokesperson for the R&D and technological dimensions of the firm gets represented in senior-level councils far more so in Japan than in Europe, which is far more so than in the United States.

Regarding the issue of control of R&D, the United States is distinguishing itself from Japan and Europe by the rush to shift control of research and development down to the bottom levels of the firm, that is, to the business unit. This will produce shorter response times for research and development, but leaving out the longer-range research activities by focusing on the near term. This trend is cyclical, and will likely reverse eventually, but not soon enough to aviod damaging corporate technological strength. For research, the trend in Japan is toward increased corporate-level control of research, counter to the U.S. movement.

The Japanese are increasing the amount of corporate–level strategic guidance of research content. This is a very significant distinction between Japanese companies and European and American firms. I believe that American companies do <u>not</u> understand this countertrend that is now taking place in Japan.

Regarding Japanese appreciation of academic research and education, our study documented four dimensions of university linkage (collaborative research efforts, obtaining innovative ideas, determining technology trends, training corporate personnel). It is clear from our findings that the Japanese companies are far more active and have higher levels of appreciation of the role of the university. Japanese firms see the universities as very productive. However, it is not the Japanese universities they see making such contributions to them, it is the American universities the Japanese firms see making contributions to them.

My final comment focuses on learning from Japan. Both American firms and Japanese firms seem to be doing equally well in listening to their customers. This has been a major change in U.S. companies during the last three to five years. Thus, we have already learned from the Japanese, and we have implemented major changes in this one area. If U.S. firms can learn from their Japanese counterparts in such a short time about understanding the voice of customer, then I believe American firms can also learn in some of these other areas.

The question I asked earlier to Kodama and Branscomb about whether they encountered resistance in telling people "yet another story of excellence in Japanese management" is also a story of my own experience. Whenever I try to convey these data to American executives in particular, they are very resistant to hearing the information and to understanding the strategic implications of it. The resistance is high.

AUDIENCE DISCUSSION

After the panel of discussants, Gerald D. Sullivan, the moderator, opened the discussion to the other speakers and to the audience.

Lewis M. Branscomb questioned Prof. Roberts regarding the issue of executives serving on corporate boards. Had his survey looked at the "management committees" as well as "boards of directors." Prof. Roberts responded that the survey answers were comingled: the question referred to either the board of directors or to the main management committees. Respondents had an opportunity to indicate whether their chief technical people were sitting on "main management boards." For the Americans, only 25% said "either."

Prof. Roberts noted that it was three generations before Branscomb's tenure (as chief technical officer) at IBM that the corporate vice president and chief scientist last served on the Board of Directors at IBM. And from a similar time frame, it was also three generations ago that the head of R&D at 3M was a board member. Since then, corporate "push-down" and "push-out" has excluded the senior technical officer, even in such great technological firms as IBM and 3M."

However, Branscomb continued, "The reason I asked the question was that I, and the two [vice presidents for research and development] who succeeded me, did serve on the Management Committee, which is where the real business strategy and discussions take place. Moreover, at IBM, "management did not tell its board anything."

Jordan J. Baruch raised a question about the incentives of the private power companies in Japan. He noted Prof. Kodama's slides showed that the Tokyo Electric Power Company (TEPCO) produced some thirty times as many kilowatt-hours of electricity as did the Boston Edison Company. Yet TEPCO's capitalization is only about six times that of Boston Edison. This implies a high rate of technological change in the delivery of electric power in Japan. One must recognize, he said, that in the U.S. the rates and profits allowed by public utility commissions to American power companies are determined by capital investment, rather than by how well they do their job. He asked how the rate and profit regulation is done in Japan. Prof. Kodama replied that prices are high and are regulated on a different basis, taking account of costs and the technology being used.

Guy Stever observed that one speaker said that "Japan has a very serious problem," and then another speaker said, "The U.S. has a very serious problem." He noted that Martha Harris spoke of two different cultures, and they are proceeding on different courses with the hope that they will cross somewhere. There is also a possibility with different courses that they can <u>diverge</u>. Stever asked if there was any evidence at present whether they are converging or diverging.

George Heaton, referring to an earlier speaker, observed that the function of management in Japan can be characterized as a means to create more entropy, whereas in the United States the purpose of management is to reduce the amount of entropy. It seemed to him that both societies are now moving in a similar direction. For example, regarding the practice of team work as opposed to individual effort, the U.S. seems to be emphasizing teamwork more, and the Japanese have been emphasizing individual creativity less. He agrees with the comment about converging trends.

Martha Harris added that she could see evidence that the U.S. and Japanese R&D systems are coming closer. Talking with younger Americans who have spent time in Japanese laboratories, studied in Japan, and speak Japanese fluently, she has become more optimistic. They can truly operate internationally, just like the many Japanese engineers who come to this country. And at the same time, there are some significant structural differences by which the respective innovation systems and market systems currently are organized. These differences are not likely to disappear soon. The key is how to structure interdependence at the mico level, —in other words, how participants in U.S.-Japan joint ventures or research exchanges are able to interact so that both sides benefit, and so that we can show, at the more macro, national level, how such interactions can indeed have mutual benefit.

Gerald Sullivan, the moderator, concluded the session by pointing out to the audience about the difference between industries and cautioned against generalizations. He noted that most of the morning discussions focused on the Japanese electronics industry. In 1983, he was with a group of high executives who went over to Japan and met with the boards of several electronics firms and aerospace firms. What they found was that the Japanese electronics firms were quite aggressive. They knew that they were World Class competitors, and were prepared to meet on a tit–for–tat basis. By contrast the aircraft industry firms seemed to be standing and waiting for the next "handout" from their government or from the United States (under the U.S.–Japan defense aircraft co–production agreement). The differences in management style, in strategy and in practices were like night and day!"

He closed by announcing that the symposium would resume in the afternoon with a panel of four university groups who are now engaged in programs designed to develop a comprehensive understanding of Japanese industry and technology management. Each school is focused on different industries: aerospace, automotive, electronics, and the environment. "Instead of producing more generalizations about what Japan does differently from the U.S., these university programs will identify those "industry–specific" characteristics associated with technology management and will describe what is needed by American scientists and engineers to function effectively in the Japanese high–tech culture.

The morning session was adjourned.

BIOGRAPHICAL SKETCHES

Morning Session

H. GUYFORD STEVER

Ph.D. Physics – California Institute of Technology (1941)

Commissioner, Carnegie Commission on Science, Technology, and Government. Formerly served as Foreign Secretary of the National Academy of Engineering; President, Carnegie Mellon University; Professor, MIT; Chief Scientist, U.S. Air Force; Director, National Science Foundation; White House Science Advisor to the President; Member: National Academy of Sciences; National Academy of Engineering; Academy of Engineering of Japan (Foreign Associate).

KANEICHIRO IMAI

Doctor of Engineering – Ministry of Education (1958)
Mechanical Engineering Course Graduate – University of Tokyo (1941)

Vice President of Japan Society for Engineering Education; Professor Emeritus of Nihon University; former member of Science Council of Japan (1985–91); president, Japan Society of Mechanical Engineers (1979); president, Japan Society of Quality Control (1987); member of Japan Society for Science Policy and Research Management, JSME, JSAS, TMSJ, (Japan) and ASME, ASEE, ASQC (USA).

Awarded Deming Application Prize (1976) as head of IHI Aeroengine and Space Operation; Fellow of the Royal Aeronautical Society (UK).

FUMIO KODAMA

Ph.D. in Engineering – University of Tokyo (1974)
Professor, Graduate School of Policy Science, Saitama University (1984)
He conducted research at Studiengruppe fuer Systemsforschung, Heidelberg, Germany (1967–1969); taught at Hamilton College, USA, as a Fulbright visiting associate professor (1978–1979).

His main analytical tool in policy science is system engineering. He has applied it to technological innovation and to dynamic evaluation of new policy instruments. Since July 1988, he holds joint appointment as research director of the National Institute of Science and Technology Policy (NISTP) of the Japanese Science and Technology Agency (STA).

He was a visiting professor at the John F. Kennedy School of Government, Harvard University (1991–1992) and visiting professor of mechanical engineering at Stanford University (1992–1993).

LEWIS M. BRANSCOMB

Ph.D., Physics – Harvard University (1949)
B.S., Physics – Duke University (1945)

Albert Pratt Public Service Professor and director of the Science, Technology and Public Policy Program at the John F. Kennedy School of Government, Harvard University.

His career includes positions as Vice President and Chief Scientist of IBM Corporation (1972–86); Chairman of the National Science Board (1980–1984); and former director of the National Bureau of Standards (now the National Institute of Standards and Technology) (1969–72) in Washington, D.C.

He has written extensively on information technology, science and technology policy, and management of technology. Recently, he coauthored, *Empowering Technology: Implementing a U.S. Strategy* (MIT Press, 1993) and *Japanese Innovation Strategies: Technology Support for Business Visions* (University Press of America, 1993).

He served on President Johnson's Science Advisory Committee and on President Reagan's National Productivity Advisory Committee. He is a member of the National Academy of Engineering, the National Academy of Sciences, the National Academy of Public Administration, and the National Research Council's Japan Committee.

YOSHIO NISHI

Ph.D. in Engineering – University of Tokyo, Japan
B.S., Electrical Engineering – Waseda University, Japan

Director of Research and Development Center, Hewlett–Packard, Palo Alto, CA. He joined H–P in 1986 as Director of the Silicon Process Laboratory. He also serves as consulting professor, Department of Electrical Engineering, Stanford University.

His career includes positions both in Japan and in the United States, involving research work on silicon process and semiconductors, as well as VLSI projects. Member of technical staff, Toshiba Corporation in Japan, in the Semiconductor Engineering Department, later at Toshiba R & D Center and as senior researcher at NEC–Toshiba Information Systems Laboratory.

He spent two years at Stanford University as a research associate in the Electronics Laboratories; is a fellow of the IEEE; a member of several other professional associations in the U.S. and in Japan, and served as a member of the Committee on Japan of the National Research Council.

HISASHI KOBAYASHI

Ph.D., Electrical Engineering – Princeton University (1967)
M.S., Electrical Engineering – University of Tokyo (1963)
B.S., Electrical Engineering – University of Tokyo (1961)

Sherman Fairchild Professor, School of Engineering and Applied Science, Princeton University. He was with IBM Research Division for 19 years; founding Director of IBM Japan Science Institute in Tokyo (now called IBM Tokyo Research Laboratory).

Served on Computer Research Panel, National Research Council (1985); Research Review Board, National University of Singapore; Academic Advisory Board, Advanced Systems Institute of British Columbia, Canada; advisory boards of Dept. of Electrical Engineering at Lehigh University, University of Pennsylvania; Science Advisory Committee of Stanford Research Institute (SRI). He is a Fellow of IEEE (1977).

MARTHA CALDWELL HARRIS

Ph.D. and M.A., Political Science with specialization in Asian Studies-University of Wisconsin at Madison

Director, Office of Japan Affairs, National Research Council.
She was a foreign research associate, University of Tokyo; taught at the University of Washington and George Washington University; and served as project director, U.S. Congressional Office of Technology Assessment.

GEORGE R. HEATON, JR.

J.D., Boston University (1974)
B.A., University of Pennsylvania (1969)

Adjunct professor of Management and Social Science, Worcester Polytechnic Institute and consultant to the World Resources Institute , Washington, D.C., and to the World Bank in Washington, D.C.

He was a visiting professor at Saitama University, Japan (1986–87); the first foreign scholar at Ministry of Health and Welfare (1989–90); and participated in Japan Committee on Industrial Performance (1992).

EDWARD B. ROBERTS

David Sarnoff Professor of Management of Technology,
MIT Sloan School of Management, Cambridge, Massachusetts

Co–founder, chairman of Pugh–Roberts Associates; Co–Founder and director, Medical Information Technology, Inc.; Co–founder and general partner, Zero Stage Capital and First Stage Capital Equity Funds.

He recently authored *Entrepreneurs in High Technology*, Oxford University Press (1991), winner of Association of American Publishers' award for Outstanding Book of 1991 in Business Management.

GERALD D. SULLIVAN (Moderator)

Private consultant on technological, economic, and international matters involving product potential and industrial alliances.

He has degrees in engineering and economics; worked 30 years for the Department of Defense in senior research and development positions; served as senior scientist, DARPA; on Defense Science Board's studies on Defense Industrial Cooperation with Pacific Rim nations; earlier, was member of Sec. McNamara's Systems Analysis team.

He is a fellow of the Vanderbilt U.S.–Japan Center for Technology Management, and adjunct professor of technology management at the University of Denver.

ROBERT S. CUTLER (Presider)

M.S., Management Science – Stevens Institute of Technology (1966)
B.S., Mechanical Engineering – University of Massachusetts (1955)

Adjunct professor, management of technology, Georgetown University, Center for International Business & Trade, Washington, D.C.

Formerly, senior staff associate, Program Evaluation Staff, National Science Foundation, Washington, D.C. (1973–1990); visiting Fulbright Research Fellow, University of Tokyo (1986–1987); staff to U.S.–Japan Working Group, Office of Science and Technology Policy, Executive Office of the President (1988).

Afternoon Session

U.S.-JAPAN INDUSTRY AND TECHNOLOGY

MANAGEMENT TRAINING PROGRAM

Panel Presentations

PANEL PRESENTATIONS

CLAUDE CAVENDER, Associate Director for Education, Academic and Industry Affairs, U.S. Air Force Office of Scientific Research, Washington, D.C.

JOHN CREIGHTON CAMPBELL, Director of Japan Technology Management Program, University of Michigan, Ann Arbor, Michigan.

PATRICIA E. GERCIK, Managing Director, MIT Japan Program, Cambridge, Massachusetts.

KAZUHIKO KAWAMURA, Director of the U.S.–Japan Program for Technology Management, Vanderbilt University, Nashville, Tennessee.

JAMES L. DAVIS, Co–Director of the U.S.–Japan Industry and Technology Management Training Program, University of Wisconsin, Madison, Wisconsin.

JORDAN J. BARUCH, Consultant, Washington, D.C.

JOHN MCSHEFFERTY, President, Gillette Research Institute, Gaithersburg, Maryland.

MASAZUMI SONE, Research Director, Nissan R&D, Inc., Ann Arbor, Michigan.

LEO YOUNG (Moderator), Staff Director for Basic Research Programs, Office of the Secretary of Defense, Washington, D.C.

U.S.–JAPAN INDUSTRY AND TECHNOLOGY MANAGEMENT TRAINING PROGRAM PRESENTATION

Robert S. Cutler (Presider)
Consultant Advisor
Washington, D.C.

In the fall of 1991, colleagues at four American universities began a new program to study Japanese language, culture, and technology management methods in order to learn more from their Japanese R&D counterparts. This activity, supported by grants from the Air Force Office of Scientific Research (AFOSR), is aimed at improving the productivity and global competitiveness of the U.S. industrial base. Termed the "U.S.–Japan Industry and Technology Management Training Program," its principal objectives are to enable university faculty and students to examine the cultural factors observed in Japanese industrial practice and to transfer useful technology management methods directly to American scientists, engineers, and R&D managers working in high-technology industries and in certain federal laboratories.

Policy-makers in America, both in government and in private industry, are now beginning to recognize the role of technology as an *asset* rather than an expense—as something to be *managed*. And its effective management may be an important key to international competitiveness.

But *why* has the federal government established a Japanese industry technology management program for U.S. universities? Historically, American industry has been the most successful and most experienced world–class technology innovator, what can university educators and industrial managers possibly learn from Japan, whose technologies were largely imported from the West, whose universities are considered second rate, and whose basic industries were rebuilt only 50 years ago?

Moreover, *what* is the best way to go about teaching American scientists and engineers, who are now employed in industry and in government R&D laboratories, the skills needed to enable them to work effectively with counterparts in Japan?

The afternoon session was designed to focus on three aspects of the so–called "Japanese Technology Management Mystique" and its possible applications:

(1)   What are some cultural factors observed in Japanese industrial practice that differ from those in the United States?

(2)   Why is advanced Japanese language training necessary when most of the high–technology communication is conducted in English?

(3)   How does one transfer Japanese technology management methods directly to American scientists, engineers, and managers who are working in high–tech industry and in federal laboratories?

These and similar questions will be addressed and discussed by four university program directors supported by AFOSR grants awarded in September 1991 and by a panel of three discussants from industry.

* * *

INTRODUCTORY STATEMENT

From

JEFF BINGAMAN (D-NM)
United States Senate
Washington, D.C.

The U.S.-Japan Management Training Program was established in 1991 to begin a process whereby Americans with science and engineering backgrounds could build bridges with Japanese research institutions, increasing U.S. access to critical technologies developed in Japan. I believe that educating American researchers in Japanese language and culture and helping them find positions in Japanese research institutions is a key to greater cooperation in science and technology that benefits both countries.

Fourteen (14) universities have been awarded grants under this program over the past three years. Fourteen is not enough, but it is an excellent start. The American researchers who emerge from these programs, conversant in Japanese and knowledgeable about Japanese technology management, will form the base of future cooperative research efforts between the U.S. and Japan.

The Air Force Office of Scientific Research has done an excellent job of implementing this program, and in the process has created a legacy of which it can be proud, one that I believe will have a profound effect on U.S.-Japan relations for many years to come.

OVERVIEW: U.S.–JAPAN INDUSTRY AND TECHNOLOGY MANAGEMENT TRAINING PROGRAM

Lt./Col. Claude Cavender
Associate Director
Education, Academic and Industry Affairs
Air Force Office of Scientific Research
Washington, D.C.

My name is Claude Cavender, the associate director of the Air Force Office of Scientific Research in Washington, D.C. I would like to give you a brief overview of AFOSR's U.S.–Japan Industry and Technology Management Training Program. My presentation covers the background of the program, its purpose and objectives, the grant awards that have been made during the past two years, and what we are doing to evaluate the program.

Congress first authorized the program in the FY 1991 National Defense Appropriations Act. The Senate Armed Services Committee report had requested that DOD establish this program. The Office of the Secretary of Defense, in turn, went out to the military services research funding agencies to ask who would be interested in managing this program for the Department of Defense. Thus, AFOSR took on the task; we are doing it for DOD, not for the Air Force itself.

We started with $2 million funding in FY 1991 and $9.6 million in 1992. The program solicitation called for proposals ranging from $1 million to $3 million. We made four awards in FY 1991 and four more awards in FY 1992.

You might be asking yourself: why did Congress establish this type of program, and especially, why did Congress want DOD to manage it? The rationale behind the Senate Armed Services Committee report was that the U.S. defense industrial base depends on the strength of our nation's overall industrial base—and that our economic well-being depends on more than just military strength: it depends on a strong, viable manufacturing base. Since Japan has demonstrated success in this area, it was thought that U.S. industry could benefit by learning more about the management of technology in Japan.

The initial objectives of the program are outlined in the accompanying chart. The formal request for grant proposals (the RFP) was intentionally broad; we did not prescribe exactly what we wanted. We asked the proposers to give us some good ideas. Since the principal participants of the program are university scientists, engineers, managers, and students—and the main constituency is obviously industry—I would say the umbrella goal of the entire program is to increase our general understanding of how Japan manages technology. This includes how it produces products better, cheaper, faster than we do. To understand how they do that, and to be more effective in Japan, we need to provide the participants in the program with training in the Japanese language and the culture. One cannot get inside the system [and work] effectively without understanding the language, and especially the culture within which it operates. Management of technology and culture are tightly linked together.

This DOD program also is intended to provide some practical experience in Japan for American scientists and engineers, leading to collaborative efforts with Japanese organizations in research projects, engineering development, and R&D management activities. In particular, we asked our university proposers to provide mechanisms by which employees from DOD and DOE laboratories could participate in their programs. In this way, I think that our national laboratories might become more effective in transferring to industry the technology that they have developed.

As previously mentioned, we made four grant awards in FY 1991 and four more in FY 1992. They were competitive awards. In 1991 we received 25 proposals, and the awards were made to MIT, University of Michigan, University of Wisconsin, and Vanderbilt University. These grants began about a year and a half ago. The four FY 1992 awards were made from 24 proposals received, nationwide. They went to Stanford University, University of California at Berkeley, jointly to the University of New Mexico/University of Texas at Austin, and to the University of Pittsburgh/Carnegie-Mellon University. Those have been going on for less than six months.

Below is a map of the United States that indicates the location of each university that received AFOSR awards during the first two years of this program. Geographic distribution was not a consideration in the selections. However, we now have a growing network of U.S.–Japan Centers at universities throughout the country. By the way, a requirement of our grants is that each center cooperate and work with the others to share ideas. In this way, we are developing a network of capability among leading U.S. academic institutions, which can offer the training and, hopefully, be a valuable resource for increasing our understanding in the United States of what makes Japanese technology management so successful.

Finally, I want to describe what we are doing for program evaluation. AFOSR recently contracted with the National Academy of Sciences to perform an assessment of the program. The Manufacturing Studies Board of the National Research Council and the NRC Office of Japan Affairs are responsible for this activity. They selected a multi-disciplinary committee of industry and academic experts chaired by Dr. Frederick F. Ling, president of the National Productivity Council, to perform the assessment. A panel of NRC Committee members have now visited each of the four universities that received awards in FY 1991. They are expected to produce an interim report in a few months, which will provide a summary of the first four activities funded in FY 1991, and possibly some mid–course recommendations for subsequent grantees. The overall program evaluation report is expected from NRC in Spring 1994.

I think it important that the results of the evaluation be disseminated to the principal constituents of the program—U.S. industry—and also to people within the federal government and in academia. We all need to work together to increase the industrial competitiveness of the United States.

Thank you.

Claude Cavender

SLIDES IN ORDER OF PRESENTATION

CLAUDE CAVENDER

Overview

US-Japan
Industry & Technology
Management Training

AAAS Symposium
Technology Management
in Japan

13 February 1993

Claude Cavender, Lt Col, USAF
Associate Director
Education, Academic & Industry Affairs

US-Japan Industry and Technology Management Training

Purpose

- Strength of US Defense Industrial Base Depends on Strength of Nation's Overall Industrial Base

- Japan Demonstrated Outstanding Abilities in Creative Management of Science and Technology

- US Can Benefit by Understanding Japan's Management Practices in Science, Engineering, and Manufacturing

Source: Senate Report 101-384

AFOSR

US-Japan Industry and Technology Management Training

Objectives

- **Scientists, Engineers, Managers, and Students Participate**

- **Increase Understanding of Japanese Industry and Technology Management Methods**

- **Training in Japanese Language and Culture**

- **Involvement in Japanese Research, Engineering Development, and Management Activities**

- **Provide Mechanism for Participation from DOD and DOE Laboratories**

AFOSR

1/15/93   5

US-Japan Industry and Technology Management Training

Awards

FY91

Massachusetts Institute of Technology

University of Michigan

University of Wisconsin and EAGLE/NTU

Vanderbilt University

FY92

Stanford University

University of California-Berkeley

University of New Mexico and University of Texas

University of Pittsburgh and Carnegie Mellon University

AFOSR

1/15/93   6

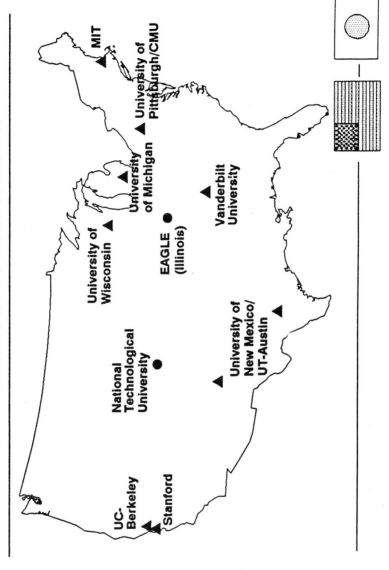

US-Japan Industry and Technology Management Training
Program Evaluation

- National Academy of Sciences Contract to Assess Program
- National Research Council (NRC)
  - Manufacturing Studies Board
  - Office of Japan Affairs
- Selected Multidisciplinary Committee of Experts
- Reports to be Published -- Intended Audience
  - US Industry
  - US Government
  - Academia

AFOSR

1/15/93 7

UNIVERSITY OF MICHIGAN
JAPAN TECHNOLOGY MANAGEMENT PROGRAM
Ann Arbor, Michigan

John Creighton Campbell
Director

Good afternoon. Our program at the University of Michigan is a joint endeavor between the College of Engineering, the School of Business Administration, the Center for Japanese Studies, and the Department of Asian Languages & Literature.

I principally represent the Center for Japanese Studies. My field is political science. Since most of our researchers are either from the business school or from the engineering school, my being from neither makes it possible for me to administer the program.

Our interpretation of the AFOSR program goals are as follows:

- To discover new knowledge about the management of technology in Japan through research

- To disseminate useful knowledge to American managers and engineers in both the public and private sectors

- To train students so that they will be able to deal with Japan more effectively in the future.

There are many ways to approach this set of objectives. The University of Michigan program has its own approach. I would like to describe it briefly to you this afternoon without going into too much detail.

It is our feeling that we simply do not know enough about how technology is managed in Japan. When intelligent people do projects on Japan and we read their reports, they seem to be learning new and surprising things quite often. Because we have not yet reached the point where we keep finding out the same things, we have put emphasis on research.

99

We have 13 separate projects going on at present. Our style in pursuing research has not been to generate brand–new projects.

To academics, particularly those in the social sciences like myself, the $2 million we received from AFOSR is a lot of money. It is more than we have ever had at Michigan for a two–year project in Japanese studies. But one finds quickly, particularly among colleagues from the engineering school or from the business school, that the amount does not go far if one is doing research.

In fact, we contacted a variety of some of the better scholars on campus and found out their research interests. We then identified the projects in which there seemed to be an interesting Japan–related aspect to what they were currently pursuing, even if they had not paid much attention to Japan in the past. To some extent, we bought into their ongoing research projects. For the most part, our funds enabled them to add Japan to projects that they had already intellectually defined. That is the way we started.

The selection was not random, since we did have three major themes. The first was "Product Design in the Automotive Industry." Located as we are in the Detroit area, there is lot of good research going on at the University of Michigan on automobiles. And, as it happens, it is the automobile industry that probably has done the most over the past 10 years to learn from the Japanese industry, for obvious reasons.

Four research projects focus on the design process—for example, concurrent engineering, how it works, and how the Japanese have managed to make cross–functional teams that also operate more effectively than ours do. Note, however, that U.S. auto makers have become better at it over the past five years or so. The implications of concurrent engineering for the intellectual process of design are that one has to think differently about how to narrow alternatives and make decisions in the design process. Another project is how suppliers and manufacturers work together during product design phases.

These are all projects by researchers who have done similar work on U.S. industries and are now seriously looking at Japan for the first time.

Our second topic area is "Globalization." It has to do with the way the Japanese manage technology and how it has been affected by globalization, by the way Japanese firms are opening up more widely to other national cultures. For example, there is a project on technology transfer across national borders, such as to the United States, being done within our business school, and a project on technology–based joint ventures and what makes them work or not work.

The third area is "Cross–Cultural Learning." The major project of the three again has to do with the automotive industry. It is being carried out by the Office for the Study of Automotive Transport at the University of Michigan. That work is based on the logic that, of all U.S. industries, the American auto industry has worked the hardest on learning from Japan. They have learned some good lessons, they have learned some not so good lessons, and they have not learned things they probably should. The purpose of this project, then, is to look at barriers to effective learning, the facilitators of learning, how people define what is to be learned, and how all that has changed over the past 12 or 13 years.

The results of the research projects will be disseminated via several channels. We have a seminar among the researchers every three months or so. The research findings will be published in normal academic fashion as working papers and as articles for journals. In addition, we are planning to hold a conference in July 1993 to prepare a book on management of technology and related Japan topics.

However, dissemination to other scholars is not the primary function of the Japan Technology Management Program. It is clear from the legislation and from the intentions of the people who started the program that bringing the information to today's managers and engineers in the United States cannot just rely on scholars doing studies and publishing in their journals. Actually, one way things get disseminated is via faculty consulting. All of our faculty members consult with industry; one works even more directly with the American Supplier Institute and the Industrial Technology Institute, which do a lot of industrial training.

We have found that the best medium for communicating contemporary working knowledge to industry people is the "short course." This can range in duration from two days to a couple of weeks. It typically involves an audience of between 30 and 100 people and is done in classroom, instructional mode. We also include our materials in the Executive Management Seminars offered by the business school to senior-level people.

The point is to get the material about Japan into a form that people in industry can use to find solutions to problems which they can recognize. Effective presenters are often those U.S. managers who have been working to apply Japanese techniques in their own factories. These people are often more convincing than academics when talking about how it is done in Japan. We plan to target similar training for government laboratories, like Wright-Patterson AFB and the Army Tank Command.

The final program area is "Student Training," mostly on our campus. The most important part is the language training. If our student engineers and technologically-minded business administration people are to be able to deal better with Japan in the future, they have to speak good Japanese.

There are many ways of teaching Japanese to engineers and to managers. I think it needs a lot of experimentation. Some schools have taken the approach of taking engineering students or even people already in the field, who know no Japanese and figuring out ways to get them to learn enough Japanese to be useful. That is an excellent thing to do.

However, our approach is different: we are working on the advanced end and have devised two new courses on technical Japanese at the third- and fourth-year levels. One needs the equivalent of two years of basic Japanese to enter the third-year course. It includes basic Japanese but also introduces technical vocabulary and some technical situations. The fourth level deals more seriously with training in engineering and science. These are experimental courses; the fourth-year course is being taught for the first time this year.

Scholarships are necessary for students in this new area. We are asking students to take on extra work. It takes a lot of extra time, particularly in engineering. Students need incentives and some way to offset the added costs. We are trying to do this through scholarships. For example, we offer a summer language scholarship for taking Japanese language at the University of Michigan. Or we can send them to Japan to learn Japanese. This program is open to engineering undergraduates and graduates and to technology–oriented business students, including students outside of Michigan. We are also offering a number of fellowships (tuition plus stipend) for graduate students who are willing to add an extra year of work in Japanese language and area studies.

Our internship program is just beginning. We do not have enough engineering students who speak good Japanese yet. These are summer internships for two or three months working on a project within a Japanese firm. We hope to expand this activity in the future to full–year internships. We are just getting into this with engineers this summer.

Regarding all these student training programs, in the long run it comes down to changing the culture. It is clear that at many university business schools, starting about 10 years ago, there has been a real cultural change about Japan. At the engineering school at Michigan, we are just starting. Over time, if we can build this into the structure of the university—the technical Japanese language courses, the certificate programs, the Japan internships—it could become a normal part of the university curriculum, rather than something added by outside grant funds.

The plans for next year are to expand our student programs along the lines of what we are doing now. In our research effort, we will focus more tightly on manufacturing *per se*. Although this would not be original research on how it is done in Japan, the work will detail the ways such Japanese methods have been incorporated by U.S. firms and how the methods fit together. We then expect to take this research and to integrate it more closely than we have before in our continuing education programs.

Thank you.

John Campbell

MASSACHUSETTS INSTITUTE OF TECHNOLOGY
MIT JAPAN PROGRAM
Cambridge, Massachusetts

Patricia E. Gercik
Managing Director

My name is Pat Gercik. I am managing director of the MIT Japan Program. This afternoon I would like to give you a little history of our program, because it describes the mechanism for what we are now doing.

The AFOSR grant has changed the program in ways we never imagined. It enabled us to go out to government laboratories and to corporations, and also to develop several new training programs and an interactive video software package that could really get specific information about Japan to the American public.

However, the philosophy of the MIT Japan Program remains constant. And I think that is what drives the program and gives it the energy needed to move forward in the 1990s.

The MIT Japan Program was founded in 1981 by Professor Richard Samuels as a contribution to redressing the perceived science and technology gap between the United States and Japan. His vision was to create a cadre of Americans who could read, write, and speak Japanese, were knowledgeable about Japanese social and business culture, and were technically trained. To date, more than 300 interns have been through the program.

As a matter of fact, when I first came to MIT and tried to get professors here to cooperate with the program, they said, "Pat, it [new technology] only happens here! Don't worry." That perception has really changed in America, and particularly now at MIT.

The MIT Japan Program now is the largest, most comprehensive center of applied Japanese studies in the United States. It is designed to attract every level of student and carry them through their academic career. The three pillars of the MIT Japan Program are education, research, and public awareness. Educational activities include internships in Japanese laboratories, language training, on-site training modules, creating educational materials, providing fellowships and grants, and offering seminars and symposia. The interns are trained in Japanese and they attend seminars and retreats on Japanese business and the R&D environment. The MIT Japan program provides two intensive summer programs for lower-level language training, and technical Japanese related to material science and electrical engineering at the upper level. In addition, we provide on-site training at American laboratories with customized cases, games, and simulations including interactive CD–ROM case studies.

Research through program sponsorship covers topics such as intellectual property rights, collaborative R&D, business–government relations, Japanese energy policy, U.S.-Japan competition in aerospace, globalization of R&D, Japanese manufacturing, Japanese political economy, U.S.-Japan defense technology cooperation, and organization of basic and applied research.

The MIT Japan Program is also engaged in activities designed to increase the general awareness of Americans about Japanese science and technology issues. To do this, we publish a newsletter on U.S.-Japan science and technology issues, maintain databases on Japanese science and technology, conduct seminars and symposia, provide access to Japanese science and technology information, conduct crisis simulation exercises, and sponsor certain cultural events. In particular, the *MIT Japan Science and Technology Newsletter* provides a resource on Japanese technology management to the MIT faculty, Japan researchers, program interns, libraries, Japanese R&D labs near MIT, corporate sponsors, and government sponsors. As such, it fills the gap between what is published in the U.S. and in Japan and it provides analysis rather than just information.

An expanded program of public seminars and symposia is now being developed to cover topics of current interest, such as issues of U.S.-Japan defense technology collaboration, market access, competitive strategies for U.S. corporations, and information about how to establish operations in Japan. Other public-awareness activities include a crisis simulation game and the sponsoring of various Japanese cultural events for students on campus.

In summary, I believe that the success of the MIT Japan Program can be attributed to a strong vision, thorough language and culture training, the development of camaraderie among the students, strong contact with both U.S. and Japanese companies, and the follow through with students both during the program and after they have completed it.

Thank you.

Patricia Gercik

SLIDES IN ORDER OF PRESENTATION

PATRICIA GERCIK

THE MIT JAPAN PROGRAM

- The MIT Japan Program was created in 1981 to redress the "Science and Technology Gap" between the United States and Japan.

- The MIT Japan Program is the largest, most comprehensive center of applied Japanese studies in the United States.

CORE ACTIVITIES

The 3 pillars of the MIT Japan Program are:

- Education
- Research
- Public Awareness

EDUCATION

Educational Activities

- Internships in Japanese Laboratories
- Language Training
- On-Site Training Modules
- Creating Educational Materials
- Fellowships & Grants
- Seminars & Symposia

Patricia E. Gercik

INTERNSHIPS

The Internship Program

- Intern candidates are drawn from the MIT student body and US government laboratories.

- Interns are trained in Japanese language.

- Interns attend seminars and retreats on Japanese business, R&D environment and culture.

- Interns are placed in Japanese corporate, government and academic laboratories.

112

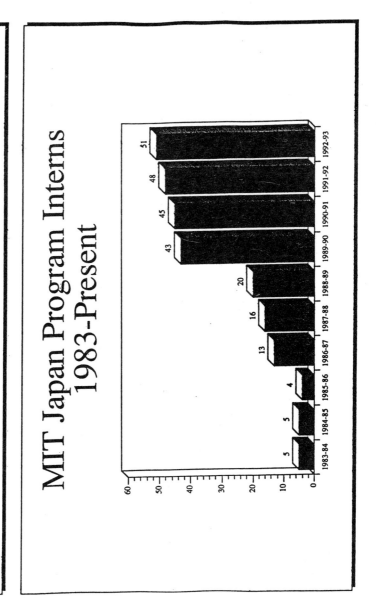

INTERNSHIPS

MIT Japan Program Interns
1983-Present

PUBLIC AWARENESS

The MIT Japan Program is engaged in many activities designed to increase awareness of US-Japan S&T issues. These include:

- Newsletters on US-Japan S&T issues
- Databases on Japanese S&T
- Seminars and Symposia
- Access to Japanese S&T information
- Crisis simulation exercises
- Cultural Events

CORPORATE SPONSORS

The **MIT Japan Program** currently has **17** US Corporate Sponsors:

- AT&T
- Cognex Corporation
- Digital Equipment Corporation
- Dow Chemical
- Eastman Kodak
- Ford Motor Company
- Fusion Systems
- IBM
- Monsanto Chemical

- Motorola
- Procter & Gamble
- Teradyne
- 3M
- The Timken Company
- Trimble Navigation
- United Technologies
- Xerox

VANDERBILT UNIVERSITY
U.S.–JAPAN PROGRAM FOR TECHNOLOGY MANAGEMENT
Nashville, Tennessee

Kazuhiko Kawamura
Director

Good afternoon. I am Kazuhiko Kawamura, Professor of Electrical Engineering and Director of the U.S.–Japan Center for Technology Management at Vanderbilt University. First I would like to introduce our program—its structure and approach—then give a few examples of what the students and faculty have been doing during our first year of the AFOSR grant, and finally present two initiatives planned for next year. Some background information follows.

The Vanderbilt program in U.S.–Japan technology management was designed to achieve two twin goals: (1) educating students and faculty to transfer useful features of Japanese technology management methods to the United States, and (2) encouraging more American scientists, engineers, and managers to work effectively with Japanese counterparts in order to increase U.S. access to Japanese research and technology through hands–on experience in Japan.

The AFOSR grant program is run under the joint auspices of the School of Engineering, the Owen Graduate School of Management, the Institute for Public Policy Studies, and the U.S.–Japan Center.

Main Program Components

This has been the first year of our program. Let me describe its three major components:

- Graduate Degree program

- Non–degree program

- Research and Internship program.

115

A Master of Engineering degree in Management of Technology with specialization in Japanese technology management was recently established, and the existing Master of Science and Interdisciplinary Ph.D. degree programs were strengthened. The program also sponsors faculty research, a non-degree program, student internships, and Japanese language training. In addition, collaborative research has been started with three Japanese organizations: the National Institute for Science and Technology Policy (NISTP), Tokyo Denki University, and Saitama University. The focuses are on human resource management, new product development, TQM toolset/software, innovation in small and medium–sized manufacturing firms, issues in intelligent manufacturing systems, and transfer of dual-use technology.

The non-degree programs include seminars, symposia, short courses, and distance education. Also, on-site symposia were conducted during the past two years for DOD, NASA, DOE, and the Air Force Institute of Technology (AFIT). Some examples are given below.

Our internship program is open to undergraduate and graduate students, as well as government and industry scientists, engineers, and managers. They can spend a summer working with counterparts in a Japanese government or industrial laboratory doing work at NISTP. In June 1992, nine student interns went to Japan.

Language training and orientation includes basic and beginning technical Japanese and instruction on history, culture, and business practices. Since its inception, the program has progressed and expanded significantly in all of these areas. In addition, we are engaged in joint programs with the University of New Mexico, the University of Texas at Austin, and the University of Pittsburgh.

Our second year's projects will focus on research toward environmentally sound manufacturing, production of a distance education course (with broadcasting scheduled for June 1993), and additional student internships in Japan.

Recent Student Activities

Let me to give examples of what our students are doing. Master's degree candidate Sara Baylor became the first full-time student in our graduate degree program in the fall of 1991. She is working under the faculty guidance of Prof. Gary Scudder of Vanderbilt's Owen Graduate School of Management. Both Ms. Baylor and Prof. Scudder visited Japan in the summer of 1992, the former as an intern in a Japanese laboratory (as a National Science Foundation fellow) and the latter as a researcher investigating the timing and quality of new product development in Japan.

In January 1992, a second student, Richard Lane, began the master's program. Mr. Lane began his studies while working as an engineer with the Saturn Corporation in Springfield, Tennessee, and has transferred to Detroit, but he is continuing his studies at Vanderbilt. He visited Japan as a program intern and studied robotics and manufacturing.

Marc Mueller and Lauren Heffelman became full time master's degree students and Michael Leahy began Ph.D. studies under the program's auspices in September 1992. Marc comes to Vanderbilt from officer status in the Navy's nuclear power program, and Lauren is with the U.S. Army Corps of Engineers. Several part-time industry Ph.D. students are also enrolled.

A new course, "Current U.S.-Japanese Relations," was developed by Program Co-Director James E. Auer. It was taught to 58 master's and Ph.D. students at Owen Graduate School of Management, as part of the MBA curriculum.

For the Spring 1992 term, Prof. Kazuhiko Kawamura and Ph.D. candidate Dona Mularkey developed and taught a special course on "Japanese Technology Management Practices" for our 10 graduate students. The class was held on Saturday mornings to accommodate those students who are also employed in government and industry. In fact, one student traveled from Huntsville and one from Wright-Patterson Air Force Base just to attend the class. Four of these students also traveled to Japan as program interns this past summer.

Guest Lecture Series

Visiting lecturers have been incorporated into the seminars of the Vanderbilt U.S.–Japan Program. Last year, Prof. Fumio Kodama, who spoke here this morning, was a visiting professor at Harvard University. He presented a guest lecture at our Saturday seminar on Japanese high–technology management. Prof. Fujio Niwa of Tsukuba University and NISTP, was visiting in the U.S. before attending a conference in Florida. He agreed to present a guest lecture on the Japanese science and technology indicator system to our students.

In addition, Ms. Fumi Hasegawa, a researcher at Fujitsu, lectured on her work on aids for the handicapped and elderly, addressing the application of technology to meet social needs in Japan. Also, today's symposium organizer, Robert S. Cutler, a former Fulbright Fellow in Japan, lectured on U.S. and Japanese technology transfer practices. Dr. Michiyuki Uenohara, Executive Advisor to NEC Corporation, visited Vanderbilt enroute to the NEC Research Institute in Princeton, New Jersey, where he is chairman of the Board of Trustees. His talk was titled "A Management View of Japanese Corporate R&D." (and is reprinted with permission in the Appendix).

Several Japan summer research and internship missions were conducted by six Vanderbilt faculty and 10 graduate students who traveled to Japan for two- and three-week periods during June and August 1992. The purpose was to visit several Japanese corporations, government agencies, and universities to discuss future internships and possible collaborative research. They were housed at Meiji Gakuin University in Tokyo. In connection with the travel missions, an intensive Japanese language course was held just before leaving for Japan. Six faculty and 10 students above took part.

U.S.–Japan Center Activities

The U.S.–Japan Center for Technology Management has been established as a joint venture of the Vanderbilt School of Engineering and the Institute of Public Policy Studies to administer the program's non–degree and research agendas. The Center conducted a number of these activities during the first year.

A paper titled "Transfer of Dual–Use Technology from Japan through Reciprocal Equity Investments," co–authored by Dona Mularkey, Jim Auer, and Kazuhiko Kawamura, was delivered at the Portland International Conference on Management of Engineering Technology (PICMET '91) by Dona Mularkey in October 1991.

A kickoff symposium, "Comparisons of U.S.–Japan Management of Technology," attended by 75 senior government, business, and academic leaders, was held at Vanderbilt on November 4, 1991.

In January 1992, the Center published the report of its Second U.S.–Japan Technology Forum, a meeting of some 20 senior American and Japanese businessmen who met in Nashville in May 1991 to discuss dual–use technology transfer and impediments thereto.

A symposium, "Japanese Management Methods: Opportunities for America?" was held at the Space and Rocket Center in Huntsville, Alabama, on February 5, 1992, and co–sponsored by the National Management Association. An audience of 85 government and industry representatives attended.

A special two–day seminar, "Understanding Japan: Profitable Opportunities for Tennessee Businesses," was co–sponsored by the State of Tennessee's Department of Economic and Community Development and by the Japan External Trade Organization (JETRO). More than 100 members of the local business community participated. This seminar was as useful to our student Japan summer interns as it was to would–be Tennessee business exporters.

Professor John Bourne completed a research report titled "Quality Tools: A Study of Japanese and Other Tools and Methods for the Improvement of Quality." This research was jointly sponsored by the U.S.–Japan Program and Northern Telecom, Inc.

Second Year Initiative

In addition to continuing the degree and non–degree elements, we are planning several efforts based on current collaborative R&D going on between the United States and Japan. This activity will focus on the synergistic relationship between protecting the environment and developing advanced manufacturing technologies.

Summary

Advanced technologies are a critical factor in the well–being of national economies and in national security. Successful management and implementation of high technologies require careful attention, not only to scientific and engineering capabilities, but also to human resources, team work, and strategic planning in rapidly changing environments.

The Vanderbilt Program in U.S.–Japan Industry and Technology Management started from a near–zero base and fostered a number of new degree and non–degree program activities, which have been well received on the Vanderbilt campus and throughout the Southeast U.S. Support from the Japanese government and industry has also been strong. We expect to continue to educate and to prepare American engineers and technology managers for mutually productive and rewarding work experiences in Japan.

Thank you.

Kazuhiko Kawamura

UNIVERSITY OF WISCONSIN
U.S.-JAPAN INDUSTRY AND TECHNOLOGY MANAGEMENT TRAINING PROGRAM
Madison, Wisconsin

James L. Davis
Co-Director

My name is Jim Davis, Assistant Professor of Technical Japanese, Department of Engineering Professional Development, and Co-Director of the U.S.-Japan Industry and Technology Management Training Program at the University of Wisconsin at Madison. By way of introduction, let me first provide some background, then describe the other organizations that constitute our AFOSR grant program, and finally I will detail some examples of the approach that we are taking to implement our engineering, management, and language training activities.

Introduction

In September 1991, the University of Wisconsin-Madison received a grant from the Air Force Office of Scientific Research (AFOSR) to pursue the objectives of the U.S.-Japan Industry and Technology Management Training Program. To meet the four objectives of the program, Wisconsin organized a project with several elements and a number of participants.

First, in order to prepare a significant number of engineering students to work in Japan, funding was provided to the Engineering Alliance for Global Education (EAGLE)—a consortium of 13 (now 15) universities—to support training in Japanese language and culture. Second, to reach scientists and technical managers in government and at corporate laboratories, seminars about Japanese technology, technical information, and management were developed for the National Technological University (NTU). Finally, to serve both students and professionals, several activities of the Technical Japanese Program at the University of Wisconsin were expanded. Japanese language instruction is now available from Wisconsin directly and by satellite through the NTU satellite network.

123

The University of Wisconsin's Technical Japanese Program offers a distance education program to conduct classroom instruction using the various educational materials that we have created. Current program options include: (1) Advanced Technology Management programs for NTU, (2) elementary Japanese courses, (3) basic technical Japanese courses,(4) a summer translation seminar, (5) certificates in Japanese studies for engineering students, and (6) a Japanese engineering leadership program. The satellite elementary Japanese course included 73 participants at eight sites during the summer of 1992 and was conducted one hour per day, five days a week, with interactive instruction using audio conferencing. In all courses, students submit their daily homework by fax.

EAGLE Japan Program

EAGLE—the Engineering Alliance for Global Education—is a consortium of 15 major engineering schools that have banded together to provide more international educational experience for their students and new graduates. A list of universities currently affiliated with EAGLE appears below. Under this grant, EAGLE is offering training in Japanese language and culture to prepare new engineering graduates for industrial internships in Japan.

Two subcontracts have been established to implement the EAGLE programs. First, a subcontract to Rose–Hulman Institute of Technology is supporting a summer course in Japan. In the summer of 1992, 55 engineering students, selected from 78 applicants, were sent to Japan to participate in the course. Thirty-seven students were placed at Koriyama, Ibaraki Prefecture, at the Texas A&M/Koriyama campus; another 18 advanced students went to the Hokkaido International Foundation in Hakodate. In addition to language training, the students received specialized cultural training about doing business in Japan.

While the EAGLE students are on their home campuses, their targeted study of Japanese language and management methods is being supported by a second subcontract, through the EAGLE secretariat at the University of Illinois at Urbana–Champaign, to each participating university. To encourage continuing study of Japan along with busy engineering curricula, each local EAGLE program provides special support through language tutorials. In addition, the students have access to the Japan-related seminars offered via television by NTU.

Although the AFOSR funding to EAGLE formally provides only language training and orientation to engineering students as preparation for a professional experience in Japan, it is an objective of EAGLE to place students into such an experience upon graduation. We believe that direct experience by young American engineers in Japanese industry is the most effective way for them to learn about Japanese technology and management methods. As a result of their preparation in the AFOSR program, at least 28 of the 38 students graduating in 1992 have received job placements in Japanese industry. As the EAGLE program and the quality of our graduates become better known, we expect an even higher rate of placement in 1993.

National Technological University Japan Programs

The National Technological University (NTU) is an accredited, private university headquartered in Fort Collins, Colorado, which delivers quality graduate degree courses, seminars, and video-teleconferences nationally by satellite. More than 40 of the nation's leading universities contribute to NTU's course offerings. Currently, 430 sites in 130 different organizations participate in the NTU Network. In addition to most major corporations and research universities, many Department of Defense (DOD) and Department of Energy (DOE) laboratories are now receiving NTU programs. With AFOSR funding, NTU has included 24 additional DoD and DoE sites as subscribers.

NTU is now offering several Japan-related programs. One is the monthly series of Advanced Technology and Management (ATM) programs, which cover aspects of Japanese culture, business management, and technology development. The series began in May 1992, and by December the total number of participating sites for the first eight programs was 302, with a total attendance of 3,904 people. Summaries of course evaluations show that the ATM programs have generally received above-average ratings.

125

A recent ATM program video offering was a one–day short course titled "The Structure of Research in Japan," broadcast via NTU satellite on November 20, 1992. This course was developed to provide views of the Japanese research system from people with a technical background who have observed the planning and execution of Japanese research projects from the inside.

NTU also broadcasts Japanese language courses to commercial sites and government labs. During the summer of 1992, 73 students enrolled in an eight–week course in elementary Japanese that originated at the University of Wisconsin. The course was broadcast via the NTU satellite to four DOD sites and four corporate sites.

Finally, NTU has broadened its "Management of Technology" (MOT) master's degree program to include Japanese technology management material. In addition, in May 1993, they are planning to take 80 graduate students and faculty for a 10–day study mission to Japan. The director of the MOT program, Prof. Alden S. Bean, has added considerable material on proven Japanese management practices. Additional Japanese examples and case studies are also being incorporated into the NTU/MOT courses. For example, distinguished Japanese faculty are invited as guest lecturers at the NTU trimester residencies.

The Technical Japanese Program

The University of Wisconsin's interest and capability in Japan and technical Japanese language tie the various components the program together. The primary goal of the Technical Japanese Program for the past 10 years has been to help scientists and engineers obtain—from Japanese books, technical articles, and documents—the important information they need to carry out their work effectively. In recent years, the scope of the program has also encompassed methods for accessing Japanese technical information and technical communication in Japanese.

A one–day short course titled "Monitoring Japanese Technical Information" was broadcast via satellite by NTU in June 1992. This course was designed to aid participants in developing their own strategy for monitoring technical developments in the government, academic, and corporate sectors in Japan. The course introduced participants to

126

Japanese technical journals, the Japanese patent system, domestic conferences, and other sources of information emanating from Japan. Scientific and technical databases, in Japanese and in English, were evaluated. The course was prepared by the Department of Engineering Professional Development at the University of Wisconsin–Madison.

The eight–week language course "Elementary Japanese" was broadcast via NTU satellite, June through August 1992. This course, to teach of reading, writing, listening and speaking in Japanese to persons who have never studied Japanese before, was prepared by the Department of Engineering Professional Development at UW–M.

A basic sequence of three semester–long technical Japanese language courses is being offered via NTU satellite broadcast and audiographic teleconferencing from September 1992 through August 1993. The basic sequence introduces a technical person who has never studied Japanese before to enough grammar and Japanese characters (kanji) for that person to read documents in his/her field with the aid of a dictionary.

To reach the greatest number of participants possible, the basic sequence is currently being offered live via satellite broadcast and via audiographics. (Audiographic teleconferencing involves a computer network for display of text and graphics, coupled with an audio conference.) Participation in the live class is strongly encouraged, but students off campus who are not able to participate in the live class view videotapes of the NTU satellite broadcasts.

Summary

The Wisconsin program has made a good start this year in pursuing the AFOSR program objectives. Especially impressive are the numbers of participants being reached: 55 EAGLE engineering students studied in Japan; 64 government employees completed a course in elementary Japanese (of whom 24 subsequently enrolled in a technical Japanese course); 11 government laboratory personnel enrolled in the NTU Management of Technology master's degree program; and hundreds of technical professionals in government laboratories, corporations, and universities have monitored each of the NTU satellite–broadcast, Advanced Technology and Management Programs.

Thank you.

James L. Davis

Engineering Alliance for Global Education (EAGLE) Consortium

Cornell University
Georgia Institute of Technology
Lehigh University
North Carolina State University
Rose-Hulman Institute of Technology
State University of New York-Buffalo
Temple University
Texas A & M University
University of California at Berkeley
University of Illinois at Urbana-Champaign
University of Michigan
University of New Mexico
University of Texas at Austin
University of Wisconsin-Madison
Vanderbilt University

---

15 Participating Engineering Schools

DISCUSSANT

Jordan J. Baruch
Consultant
Washington, D.C.

My name is Jordan Baruch. I am a consultant in Washington, D.C., and for many years served as an MIT faculty member in electrical engineering and a faculty member of the Harvard Business School. I summed up my academic life at Dartmouth College, where I was simultaneously a professor of both engineering and of business administration. During President Carter's administration, I was Assistant Secretary of Commerce for Science and Technology. Because I was deeply involved in both technology and in management, I became interested in what made some companies successful doing things one way, while other companies were successful doing things in other ways.

For example, some years ago I studied the Ampex Corporation, which had developed videotape recording—a fantastic technological development. During the same period, the Japanese also took hold of this new technology—the VCR—and ended up with one in an awful lot of homes. I began to ask myself, "*Why?*" And today I got a lot of answers to that question.

I would like to be somewhat controversial. We have talked about what we can learn from Japan in American industry. We have talked some about what Japanese industry can learn from America. And we have talked a great deal about the drive to competitiveness. I would like to widen the focal point a bit.

When I was a youngster, I learned that if two people pulled on the ends of a rope, each with 100 pounds, there would be 100 pounds of tension on the rope. But if they tied it to something and both pulled on the same end, there would be 200 pounds of tension on the rope with no more effort than in the first case, where the two were competing and achieved only 100 pounds of tension. There is a lesson there about competition and cooperation.

I would like to review what we learned about Japan in today's sessions and then lay out an agenda (this is for Bob Cutler) for the 1995 AAAS "Science in Japan" symposium, not on technology, *per se*, but on technological change. As we get better at doing things in industry, I think we attempt to manage the process of technological change.

This morning we talked about organizational patterns. In Japanese industry we found that teams were the focus; reward systems were set up to reward teams, or reward individuals for team performance; team leaders were promoted which provided incentive for others on the team to make the team excel. In the U.S., we found quite a different situation. We found a lot of focus on inventors, small entrepreneurs, and medium-sized companies—a big push for individual accomplishment.

When we looked at training in school and early graduates in Japan, we found that the team finished the early graduates' education, indoctrinating them into "the way of the team." In the U.S., we also had industrial education of new graduates, but primarily with a focus on figuring out just where that new graduate would fit into the corporate structure and contribute most as an individual. We found that the Japanese industries encouraged incremental improvement innovation, and did it spectacularly well. Little by little they could make things better, with a very short cycle, rapid development of new product. In the U.S., we tended instead to have long product cycles, big steps in the introduction of new products, not necessarily well carried to completion. Yet, I heard us asking the question, "*How* can these two trajectories—these two different kinds of behaviors—come together, cross, and become similar?" Then I thought to myself, "Gosh, what a waste."

I would like to ask this group (the AAAS/Japan Society for Science Policy and Research Management) to start to consider, for the next joint AAAS/Japan symposium in 1995, the following question:

"How can Japan and Japanese industry, doing what it does well, work together with American industry, doing what it does well, to advance the standard of living and form of life throughout the rest of the world, including their own two countries—Japan and the U.S.—before the 21st Century?"

So rather than change their nature, I'd like us to focus on How can these two countries—these two kinds of industry—collaborate with each other by doing what each does well to increase the economic and technical output of both countries.

I have chosen to share with you the fundamental question that today's sessions have raised in my mind. And I hope that MIT and the other universities, who are now getting into the Japanese technology management culture and language and who already know the American culture, will start to discover ways they can collaborate with minimal loss of their real skills.

DISCUSSANT

John McShefferty
President
Gillette Research Institute
Washington, D.C.

My name is John McShefferty. I am president of the Gillette Research Institute, a division of the Gillette Company. I'm also a member of the Industrial Research Institute (IRI).

Some of you may not know about the IRI organization, which was established some 55 years ago with members from among the largest industrial companies in America. The current 320 members and their alternates are the corporate VPs of R&D or directors of research from different corporate divisions. We get together several times each year to discuss common issues related to the management of R&D, since our focus is on "management." I am also the chairman of IRI's International Committee. There we focus on globalization concerns, like how our companies can become more global.

I did not come today with any prepared text, just my notes and observations from the session this morning and from this afternoon. I heard from some of the discussions this morning about differences between Japanese and American companies, and I listened this afternoon about Japanese language skills and cooperative workers. I think we've learned in U.S. industry, and we are learning and changing to team-work, to "interdisciplinary" teams. We are also looking for joint ventures, working with other people, and we are interested in how can we learn from others, not only from Japan but from the rest of the world, as well.

I would ask the speakers from this afternoon's session: How do you develop your programs? You have described your universities' new Japan industry and technology management programs—and they are all very interesting—but how do you get the actual "inputs" from the

133

industrial people? I know that some companies now support your activities, but how do you communicate with them and how often do you communicate with them? How do you understand what industry needs in America?

Patricia Gercik, the director of the MIT Japan Program, responded that she makes it her business to go out to the companies associated with the MIT program. And she said that she works closely with them to identify **what** they need to know and then put together the training materials specifically addressing those needs. "That is one way to work with them. Another way is to find out which part of a large company might find the information we have at a certain meeting to be useful to them, and then work very hard to get those people into a room and talk with them. So one tries to have a continuing dialogue with these companies, mostly on site at the company figuring out what they need to know. They are very outspoken about *what* they need to know." We talk about what training programs are going on, how we can add value to them. Some of these companies have very extensive training programs and we have specific ways to add value to them. We also ask them to identify the research projects they want, and we try to find out *what* they need to know. So she works in a number of ways with industrial companies.

John McShefferty: One of the things that IRI is particularly interested in is networking throughout the world. We want to find out what people in the universities and in our national labs and others are doing throughout the world in research or technology development that we can utilize to develop products, and the big problem in that is "language." I am particularly interested in the extensive programs you described at your universities on the Japanese language. And we face the same problem with the former Soviet Union, of trying to understand what is going on in Russia and in the other new republics. The biggest problem there is the language barrier. So I think this activity is a tremendous step forward.

We would welcome a lot of Japanese-proficient technical people into our organizations, if they can help us translate and also understand the culture of the country and be able to go back and find out what is going on in Japan that could help us. And we can obviously learn from one another. As a matter of fact, we have already learned much and are now making changes in what we do in America. I think the Japanese are also learning from us and making changes in what they are doing. And together we can learn from others, not only from the Japanese but also from the Europeans as well.

What I found very encouraging today is learning that there are now lots of American engineering students being trained in Japanese language and business culture. We, in industry, welcome them and would like to get some of them into our companies.

DISCUSSANT

Masazumi Sone
Research Director
Nissan R&D, Inc.
Ann Arbor, Michigan

I am director of research for Nissan R&D, Inc., located at Ann Arbor, Michigan. We are an American company which happens to be owned by the Nissan Motor Company in Japan.

One of my responsibilities since coming to the United States five years ago, is to maintain a two-way bridge with the U.S. academic and research community. In 1988, I established Nissan's International Exchange program, which supports students, post-docs, and faculty research scientists who work in Japan from one month to two years.

I have enjoyed full cooperation from each of the four university programs whose presentations we heard this afternoon. This, I think, is one of the reasons for my being here today.

Generally speaking, I think this university program is well received on both the U.S. and Japanese sides. From the Japanese viewpoint, they see that more American people are trying to understand Japanese culture and its management methods and feel basically that it is a good and essential thing to do. And I believe that Japanese companies and government organizations will continue to support and cooperate with these AFOSR grant-supported university programs.

For example, the AFOSR university program includes sending business and engineering students to Japan. This is a very important element. These people will see Japanese technology management and culture as a bigger picture. This firsthand experience provides opportunity for more Americans to improve their mutual understanding of technology management over the long term.

From the viewpoint of a Japanese organization, there are many benefits in accepting American students or company engineers as interns. Of course, there is the internationalization in Japanese organizations with the desire for some exchanges of cultural and technical knowledge. But there are also difficulties affecting Americans living in Japan, such as housing and living expense and concerns about family matters such as children's education and the spouse's job. We also consider these issues.

These university programs focus not only on technology management, but also on cultural and language proficiency. I think the key point of Japanese technology management is "people." Learning the language is important. The Japanese language is one of the most difficult languages in the world. But it is even more important to understand the Japanese culture, particularly the ways in which technology innovations are developed in the company and then used by the customers.

Moreover, I wonder whether one can apply Japanese technology management methods effectively to U.S. counterpart organizations. My reason is that the culture and the people are really quite different. And I believe that the critical element in technology management is the people.

Here are two examples of why I fear some difficulty. First, there is the practice of simultaneous engineering or "concurrent engineering," not only in the auto industry but also elsewhere. That means, let's say, that there are several departments—the design department and the experimental test department—which must work together. This should be a good thing for all to accomplish in order to develop a new product quickly. But actually, there is much negotiation and coordination of specifications and work schedules to be done among the departments; the middle managers and senior engineers expend much time and effort negotiating and coordinating. Consequently, the product development cycle takes a shorter time to introduce the product to the market.

The second is the concept of teamwork. We heard that Japanese management is "team oriented." This may be true, and it is a very effective way to develop a product. The team leader sets goals for the team, rather than goals for individuals. If, for example, a young engineer had a good idea but more senior engineers in the group did not agree, his idea would be held up or rejected. That leads to frustration and an unwillingness to offer ideas. This situation stems from tradition-al Japanese culture, where "the single nail that stands out gets hammered down." Important matters are overlooked when judging Japanese management methods from surface appearances only. For a deeper understanding, it would be better to find out from the bottom.

Finally, I think it important to continue these U.S.-Japan university programs. I hope that such activities like the U.S.-Japan technology management programs will provide a cross-cultural breakthrough to find the best solutions.

AUDIENCE DISCUSSION

Dr. Leo Young, panel moderator, asked the three discussants to comment on two broad questions: one relating to whether the practices of Japanese technology management and American technology management were actually coming closer together, or influencing one another, like U.S. companies doing more teamwork and Japanese companies beginning to appreciate the usefulness of individuality, and the second question, relating to the role of Japanese universities in the management of basic research—how it is done in each country and how such activities might connect. Are we changing basic research?

John Campbell responded by pointing out that there are a finite number of industrial problems in the world and many ways of handling them. "We emphasize differences in approach. But 80 percent of the problems are handled pretty much the same way in any advanced industrial country. Regarding the Japanese "teamwork" concept, I would agree with Mr. Sone, who characterized it as "a process of permanent and largely conflicting negotiations that goes on to make concurrent education work." We say that American firms can learn about that, but others believe it is mainly because of Japanese culture that they are so harmonious, that teamwork comes more naturally to them.   And American culture is as different to the Japanese as their culture is to us. But there is no question that the Japanese have come to the United States and to Europe, over the past hundred years, and have learned all sorts of ways of managing technology. And I believe that American companies are quite up to doing the same. The Japanese management mystique has been a dilemma.   But over the past 10 years in the Automobile industry, for example, people have been able to focus on specific areas of Japanese technology management to see how parts of it can be taken from Japan and adapted into the U.S. situation."

Patricia Gercik added that her experience working different companies indicates that they are eager to learn Japanese management practices. "And on the Japanese side, one reason given for having the

MIT student interns is that they like their researchers to work alongside foreigners, because they think that somehow creativity will rub off. The question about the Japanese universities is an interesting one. It is quite right that the universities do not do the cutting-edge research in many fields. In some fields they do, and the laboratories are excellent. However, the professors in those university laboratories are incredibly powerful, both in government and in private companies. They provide the technical information flow, in Japan, that is extremely influential and useful. And our students who work in those universities have access to that information flow and are also privy to a great many contacts and an understanding of the Japanese technology process."

Kaz Kawamura commented that Japanese educators have recently decided to look into the status of technology management curriculum in the universities in Japan and the U.S. "They are very interested in what is being done in the United States because they decided to develop similar programs in Japan. They are very systematic in their approach and are now sending people to visit various U.S. universities, I understand; also what they are interested in is how Japanese universities might do this kind of applied research. That is one current aspect of the convergence that several people discussed earlier today. The management of basic research is a particularly interesting topic. Many in the U.S. believe that the research consortium approach in Japan was a very successful idea for managing basic research. There are now several examples in the U.S., like MCC and Sematech. As a matter of fact, the trend of research consortia in the United States is going up, while the trend in Japan is going down. The ways they manage consortia in the two countries are quite different."

The university speakers' response to Dr. Young's two questions was concluded by **Jim Davis** of University of Wisconsin. He felt that the most important component is "information," and the value placed on that information. "Japanese companies and Japanese governmental organizations, people at high and at lower levels, work hard at gathering information from all sources. Information flows from the U.S. to Japan as if the Pacific ocean were a superconducting liquid. Information flows from Japan to the United States much more slowly. From my perspective, in American industry and as a university engineering professor working at Kyoto University, people in Japan view information of all kinds as an essential resource even if they don't know what they are going to do with the information. In the United States, it seems to me, that we take information for granted. And if there is any shift going on now, it may be on the part of some people in the United States recognizing that we need information, regardless of the source, and that may lead to the closing of the "gap," which I believe is happening. We are becoming similar; we are competing for the same markets, making the same products, and we want to sell to the same people. So I think that "gap" is certainly closing."

Jordan Baruch expanded on Jim Davis' "information" comment. "From an IEEE information dissemination study some years ago, I found that American engineers do not collect a lot of information. Rather, they prefer to collect a lot of information on *who* knows the particular kind information they might want. They make much use of the telephone, keep databases on people to get information from, access computer bulletin boards, and attend conferences. The Americans seem to be more interested in "access to information" than in the "collection of primary information," *per se*. And I have found that to be a big difference in the way information is handled in many countries, particularly for technical information."

A person from NASA added that it was hard to get engineers and managers in the U.S. to look at the value of information. "But if one pushes long enough, it can be done—for example, in doing research on patents. With Japanese information, however, we hit a barrier. We can't find out anyway, so we just will not look. I think these new university programs are ways for breaking that Japanese information barrier down."

John Campbell responded that many American-based Japanese technical information services had failed. There apparently was no viable market among U.S. scientists and engineers for such services. Jim Davis, added that his university program recently addressed the topic "Monitoring Japanese Technical Information." He mentioned the existence of primary and secondary sources in Japan and work done at Wisconsin to access the JICST database in English and in Japanese using a Macintosh computer.

Frank Riley, senior vice president of Bodine Corporation and a member of the NRC study team assessing the US-Japan technology management programs at grantee universities for AFOSR, asked if there were specific concerns about the nature of this AFOSR-sponsored U.S.-Japan program. Dr. Young replied that he did not believe there were any negative concerns. The program was originally initiated by Congress in 1991 and assigned to DOD, where it was very reluctantly accepted at first. But things seem to be going well at the moment.

Hideyoshi Hayashi, the science councillor at the Embassy of Japan in Washington, D.C., was pleased to hear from the last panel that such famous American universities have U.S.-Japanese technology programs. He wished to comment on the recent Japan Summer Institute program that Prof. Kawamura of Vanderbilt had described. Under a joint cooperation program at NSF and NIH, the Japanese government has invited American science and engineering students to Japan during their summer vacation period. They stay mainly in Tsukuba Science City and can do research at government and some private laboratories and learn Japanese language and culture, for about two months. 1993 is the third year of this summer internship program. From speaking with many students upon their return to the United States, he found them to have very positive feelings. Mr. Hayashi noted that such collaborative effort between U.S. universities and Japanese science and technology agencies is very important.

The afternoon session closed with the comment by Patricia Gercik of MIT that what the university programs are trying to do is not just learn Japanese science and technology, *per se*, but rather to learn a different approach to science and technology information gathering.

The session was adjourned.

*   *   *

The symposium concluded with an informal social reception for the speakers, guests, and sponsors held in the Commonwealth Room of the Sheraton-Boston Hotel, Boston, Massachusetts.

BIOGRAPHICAL SKETCHES

Afternoon Session

CLAUDE CAVENDER

M.S., Biochemistry; and B.S., Chemistry - University of Georgia

Lieutenant Colonel, U.S. Air Force; Associate Director for Education, Academic and Industry Affairs, Air Force Office of Scientific Research, Washington, D.C.; first program director, U.S.-Japan Technology Management Training Program (1991-1993).

Prior assignments include: Assistant Professor of Chemistry, U.S. Air Force Academy; staff officer, Headquarters, Strategic Air Command; member, SAC Inspector General team; base tours of duty at McConnell AFB, Goose Air Base, Labrador, Canada, and Carswell AFB, TX.

He has received the Air Force Meritorious Service Medal and the Air Force Commendation Medal.

JOHN CREIGHTON CAMPBELL

Ph.D., Political Science-East Asia Institute, Columbia University (1973)
B.A. - Columbia College (1965)

Advanced training at Interuniversity Center for Japanese Language Studies, Tokyo, Japan (1965-66)

Professor of Political Science and Director of the Japan Technology Management Program; formerly, director of East Asia Business Program (1984-1988).

His publications include: Automobile Industry and Public policy, in
The American and Japanese Auto Industries in Transition, edited by
Cole and Yakushiji (1984); *Politics and Culture in Japan*, Monograph
in Politics and Culture Series, edited by Samuel H. Barnes, Institute for
Social Research, University of Michigan, Ann Arbor (1989).

He was a member of the U.S.-Japan Automotive Industry Conference
Steering Committee (1981-1989).

PATRICIA E. GERCIK

Attended - University of Kyoto, Japan
Graduated - University of California at Berkeley (1966)
Master's degree - Tufts University (1967)

Managing Director, MIT Japan Program. She spent the first 20 years of
her life in Japan; attended two years at University of Kyoto; worked as
travel guide, facilitator of Japanese TV; taught, developed curriculum on
Japanese history and culture at Japan Society of Boston. Recent book,
On Track with the Japanese, Kokanesha Press, September 1992.

KAZUHIKO KAWAMURA

Ph.D. Electrical Engineering - University of Michigan (1971)
M.S. Electrical Engineering - University of California/Berkeley (1966)
B.S. Electrical Engineering - Waseda University, Japan (1963)

Professor, Electrical Engineering and Management of Technology, and
Director, U.S.-Japan Program in Technology Management; Associate
Director, Center for Intelligent Systems, Vanderbilt University.

He is an American citizen born in Japan. His expertise includes:
robotics, artificial intelligence, technology policy, and technology
management. He taught at the University of Michigan, Dearborn, and
Kyoto University, Japan; worked at Battelle Columbus Laboratories.

Member, AAAS Committee on Science, Engineering and Public Policy
(1984-1990).

146

JAMES L. DAVIS

Ph.D., Forestry - University of Wisconsin-Madison (1987)
M.S., Forestry - University of Wisconsin-Madison (1982)
B.S., Chemical Engineering - University of Rochester (1975)

Assistant Professor of Technical Japanese in the Department of Engineering Professional Development, University of Wisconsin-Madison. He teaches technical Japanese and has developed short courses related to Japanese science and technology since 1990.

He was a Fulbright Fellow at Kyoto University in Japan.

JORDAN J. BARUCH

Sc.D., Electrical Instrumentation - MIT (1950)
M.S., Electrical Engineering - MIT (1948)
B.S., Electrical Engineering - MIT (1947)

Consultant, with interests in the transfer of science and technology to developing countries; formerly, Assistant Secretary of Commerce for Science & Technology (1977-1981); Professor of Engineering and Professor of Business Administration, Dartmouth College (1973-1977). Member of the National Academy of Engineering.

JOHN MCSHEFFERTY

Ph.D., Medicinal Chemistry - University of Glasgow (1957)
B.Sc., Chemistry (with honors) - University of Glasgow (1953)
B.Sc., Pharmacy (with honors) - University of Glasgow (1953)

President, Gillette Research Institute, Gaithersburg, MD. Formerly, vice president, Personal Care Products, International Playtex (1977-1979); director of pharmaceutical development, Janssen R&D, Inc.; senior scientist, Ortho Pharmaceutical Corporation, Johnson & Johnson (1962-1977). After teaching chemistry at the University of Glasgow, he joined the Sterling Winthrop Research Institute, Rensselaer, NY, as a chemist.

MASAZUMI SONE

M.E., Electrical Engineering - Hokkaido University, Japan (1975)
B.E., Electrical Engineering - Hokkaido University, Japan (1973)

Director of Research, Nissan Research & Development, Inc., 1988-93. He has been responsible for Nissan's research activities in North America, including heading the Cambridge Basic Research Laboratory, organizing collaborative research with universities, and the promotion of Nissan's international exchange program (Nissan Future Technology Program).

He was a research engineer in the field of electromagnetic compatibility at Nissan's Electronics Research Laboratory, Japan (1975-82); member, Technological Strategy Planning Office at Nissan, Japan (1982-87). His assignments included: strategic planning of new technology, technology transfer, and organizational development.

Assistant Chairman, International Congress on Transportation Electronics (1988-90); member, Manufacturing Research Exchange Foundation, Japan-U.S. Cooperative Research Program (sponsored by National Research Council, 1988); and member, Board of Directors, Institute of Magnesium Technology, Quebec, Canada (1992).

LEO YOUNG (Moderator)

Doctor of Engineering - Cambridge University, England
Degrees in mathematics (1945), physics (1947), electrical engineering. Cambridge University, England

Special Assistant for Research, Office of Secretary of Defense, Washington, D.C.; involved in science and technology management; planning of basic research programs. Previously employed by the Naval Research Laboratory, Stanford Research Institute, and Westinghouse Electric Corporation.

Past president, IEEE (1980), fellow of IEEE; fellow of AAAS; honorary Doctorate of Humane Letters, Johns Hopkins University (1989).

ROBERT S. CUTLER (Presider)

M.S., Management Science - Stevens Institute of Technology (1966)
B.S., Mechanical Engineering - University of Massachusetts (1955)

Consultant Advisor, Management of Technology Program of the National Technological University; adjunct professor, management of technology, Georgetown University, Center for International Business & Trade, Washington, D.C.

Formerly, senior staff associate, Program Evaluation Staff, National Science Foundation, Washington, D.C. (1973-1990); visiting Fulbright Research Fellow, University of Tokyo (1986-1987); staff to U.S.-Japan Working Group, Office of Science and Technology Policy, Executive Office of the President (1988).

*   *   *   *   *

APPENDIX

CONTRIBUTED PAPERS

Yasutsugu Takeda

Azusa Tomiura

Michiyuki Uenohara

SYNERGETIC MANAGEMENT OF TECHNOLOGY FOR THE 21st CENTURY

Yasutsugu Takada
Executive Managing Director
Hitachi, Ltd.
Tokyo, Japan

1993 COMMENCEMENT ADDRESS
NATIONAL TECHNOLOGICAL UNIVERSITY

Mission Inn
Howey-in-the-Hills, Florida
January 9, 1993

I would like to express my gratitutde to President Lionel S. Baldwin for his invitation to participate in the formal commencement ceremony for the graduating class of NTU's Management of Technology program—at the beginning of 1993. It is indeed a great honor for me.

First of all, I would like to express my sincere congratulations to each of you graduating from the the MOT course with the Degree of Masters of Science. NTU was established, I understand, to provide a basis for further educational enrichment of future industry leaders—for people who are now working in government organizations or private companies—by learning advanced knowledge and technology management methods while at work. And that NTU is now the world's most advanced television-based university in this regard.

Reprinted with permission from *Research●Technology Management*, Nov-Dec. 1993, pp. 8-9.
Copyright 1993, Industrial Research Institute, Inc.

I was deeply moved to hear that more than 40 universities and over 100 companies are actively supporting NTU. And that NTU is utilizing advanced digital technologies for broadcasting and telecommunication. I feel that NTU's video-education system is full of "frontier spirit."

I am sure that since you are now successfully graduating from such a wonderful education program, by grasping such opportunity and pursuing your studies to receive the Master's Degree in parallel with your company work, you will perform innovative accomplishments in each of your professional situations, such as proposing a new scheme for technology development or devising new ways of determining product specifications in response to the ardent expectation of your own working environment.

The power and strength you have acquired through NTU will shortly make a major impact on industries in the United States, and around the world, which is now preparing to enter a new century. Hopefully, you will also contribute to the healthy growth of our global society.

Today, I like to talk about "Synergetic Management of Technology for the 21st Century." This is because this theme will be one of the leading concepts for the future of MOT.

By the way, before stating my expectations for the next century, let me reflect on the past 20th Century period. I imagine that future historians will call the 20thth Century, "the age of positive growth and consumption," or even "the age of chaos." It is true that during the last century, overall, the World's economy has grown about ten times. However, at the same time, this growth is associated with huge consumption of resources, including oil and minerals. Therefore, we have new global problems such as the destruction of the ozone layer or the warming up of the earth that were never thought of by people before the 19th Century. Along with the development of a World economy, expansion of north-south economic differences and tragic regional conflicts are continuing.

I hope the forthcoming century will be characterized as a wonderful age, and be remembered by future historians as "the century of reason." That is, the 21ˢᵗ Century will be the age when people all over the world will cooperate to conserve the global environment and its limited resources. However, we will have to prepare the way, beforehand, when people will cooperate in the formation of a global-scale "sense of values." Therefore, I propose here today that the guiding principle of MOT for the 21ˢᵗ Century, should be "the contribution to human progress with a deep consideration for the stability and harmony of the earth," rather than only growth through the realization of economic benefits to human beings.

The symptoms of this big change are observed as several flows in the present society already moving towards the next century. I would like to point out several such promising symptoms of change, from among the current activities of private companies, especially activities related to the management of technology for the benefit of humankind.

The **first** is the change of "value sense" underlying industrial activities. IT is quite obvious that a major shift in business policy is now being undertaken, from the sole considerations of consumers' need to those much wider social needs. Accordingly, the important R&D in industries, for example in Hitachi, are shifting to:

(1) Energy, environment, and the recycling of resources
(2) Social infra-structure (such as water supply, drainage systems, railway systems, and transportation systems)
(3) Systems related to health, medicare and human safety
(4) Electronics and software for the realization of an advanced information society

These are the macroscopic considerations of the business area. In each business area, however, we take into account the specific needs of a community with different cultural backgrounds, as exemplified by phrases such as "solution business" and "adaptive technology."

The **second** one is the change in the relationship between investments in R&D and its return. For the past 15 years, since the oil crises of the 1970's, Japanese high-technology industries, including Hitachi, have steadily increased their R&D investment. At the same time, during these years, the economic return from the business has also increased in a seemingly proportionate way. Thus, competition in R&D has started and has become over-heated, resulting in a shortening of product life-cycles. Eventually returns from the R&D investment gradually decreased. But it is impossible to cease the investment in R&D in order to stay competitive in business. Thus, the so-called "High-Tech Syndrome" has become a rather over-used phrase in Japan. Some reflections on this situation have come from among industry leaders. The importance of a "cooperative mind" became much more advocated. In other words, the necessity for effective and concentrated utilization of R&D resources has increased. Furthermore, cooperative alliances between companies all over the world are now emerging. For example, at Hitachi, we are now promoting cooperation with Texas Instruments company in semiconductors, with Hewlett-Packard in workstations, with General Electric in electric power technology, and so on. This cooperative R&D trend will be accelerated more in the future.

Third, I would like to stress that the innovation process, itself, is also changing. It is an image of the past, where some individual genius makes an invention based on one scientific principle or one single technology, in a single cultural environment, at a single location, without any help or cooperation from others. I believe this kind of "Lone-Ranger" innovation could only have happened until the beginning of the 20th Century. Today is the age when technology innovations can only be realized through the cooperation of many researchers or engineers, utilizing many scientific principles and feasible technologies. Also, information related to such technology innovation is communicated all over the world and across boundaries of different geography and different cultures. That is, we now are in the age when "many-many" people on earth will participate in realizing the innovation. However, attention must still be paid to the fact that technology innovation is not born from merely summing up linear activities, but it is brought about through an interactive "synergy effect" of people working together.

During my early days in Hitachi in 1960's and early 1970's, I was a researcher in the optoelectronics area. Therefore, I have always paid attention to the research and development process of laser technology, and the process by which its development made a contribution to human society through its applied technologies, such as optical-communication, compact disks, laser printers, and laser surgery. I think this is a typical example of the technology innovation process through the synergetic interactions among people in many kinds of disciplines and technologies, and social needs. In this field, as you know, many researchers from the United States, Europe, Russia, and Japan contributed to the realization of this technology innovation.

In the future, bio-technologies and superconducting technologies are expected to flourish. Also in these fields, larger scale synergy will be the key to a success. I am confident that researchers in South America, and the Pacific rim countries will also participate in these technology developments.

Synergetic management of technology innovation is also becoming more and more important in business activities. Innovations in a private company are mainly evolutionary. However, concurrent cooperation between the many different divisions within the company is necessary to start a new business. In the conventional style of management, starting from: marketing, product concept, determination of specifications, determination of R&D schedule, estimation for investment for production facilities, exploration of sales channels, and so on—were each performed step-by-step, one after another. Today, however, and especially in the future, these activities must be integrated, concurrently and cooperatively, into a single, unified whole process.

Recently, Hitachi started the "Strategic Business Project" system which is intended to implement the just-said concurrent management concept. This system is a new example of "matrix management," where a project is organized on a priority basis over many divisions and ranks. In this system, however, the span of the divisions and ranks is spread much wider than the previous system, where only technological divisions and ranks were involved. Now, the membership in this project spans from top executives to engineers and sales staff. In order to create new businesses for the future society, of course, a big responsibility exists for the executive director in charge of technology and R&D. However, I would like to emphasize that at the same time, those directors in charge of other business functions, such as sales, finance and accounting departments, accept responsibilities of almost equal weight.

Most technology innovations in industry are "evolutionary," rather than revolutionary. However, evolutionary innovations alone give doubtful concern for the future prosperity of the company. Industries should also contribute to revolutionary innovation. Therefore, it is necessary to consider a new financial infrastructure in order to be able to support the desired "revolutionary" technology innovation.

Today, many industries are investing in R&D as much as 10 percent of sales. Hitachi is no exception. Therefore, I would like to propose that about 1 percent of all R&D should go to strategic, long-range research. I have named such research "North Star Research." North Star Research is the tangible company investment in its long-term future, as well as having a related junction point with academia. Hitachi now has 9 corporate research laboratories with 4,500 employees in staffs, where Applied Research, Objective Fundamental Research and North Star Research are pursued.

In addition, to solidify the above described "synergetic" cooperation with academia, we established in 1989 the Hitachi Cambridge Laboratory, thought he cooperation of the University of Cambridge in the United Kingdom, and the Hitachi Dublin Laboratory, through the cooperation of Trinity College of Dublin University in Ireland. The University of Cambridge and Trinity College have been exploring fundamental researches and have been sending bright and highly capable young graduates into society.

On the other hand, Hitachi has accumulated many advanced technologies. The Hitachi laboratories at Cambridge and at Dublin are organized to achieve full synergetic cooperation with the universities.

Next I would like to talk about the Hitachi Research Visit Programme, or as we call it, HIVIPS. Since 1985, Hitachi has pursued these fellowship programs to invite young researchers with Ph.D. or M.S. degrees from many foreign countries including the United States, to visit and work in our corporate research laboratories in Japan. Up to now, more than 300 foreign researchers have worked for one or two years in Hitachi's research laboratories. I believe that this is quite significant, especially for the 21st Century world—that young researchers, with different cultural backgrounds, can experience working together cooperatively. They will contribute to the new international high-tech culture, and at the same time they will gain something fresh from Hitachi's way of synergetic management of technology.

The ultimate purpose of management of technology, I believe, is the contribution to the global society. I wish to emphasize, here today, that an increasing awareness of this type of cross-cultural cooperation among leaders of all nations is important, especially for the realization of our vision of the 21st Century as an era of reason and human progress.

I hope that my talk today will stimulate you, future leaders of American industry, who are now graduating from NTU, to recognize the importance of your personal role in making this vision a reality.

I would like to conclude my remarks by wishing each of you success in realizing the future expectations of your NTU facility, your supportive employers, and of course, yourselves.

I am proud of you all. YOU CAN DO IT!

Thank you.

Yasutsugu Takeda

PRODUCTIVITY MANAGEMENT IN JAPAN'S MANUFACTURING INDUSTRY[1]

Azusa Tomiura
Managing Director
Nippon Steel Corporation
Tokyo, Japan

Introduction

Working in Japan's iron and steel industry for over 30 years, I have been deeply involved in every stage of the Japanese iron and steel industry reconstruction since World War II. And I have engaged in promoting and managing the process of technological development for many years. Throughout my career, I have discussed the structure and characteristics of productivity of Japanese industry with a good number of professors and managing directors from other industries.

First, I will present my opinion about the improvement in productivity of Japan's manufacturing industries by using the iron and steel industry as an example. Many Japanese manufacturing industries have done the same as the Japanese iron and steel industry did. Therefore, I believe that my presentation about my experience in the iron and steel industry would be applicable to other Japanese manufacturing industries, as well.

Japanese Iron and Steel Industry Productivity

Labor productivity in Japan has increased over nine-fold in the past three decades. In the 1960s, improvement in productivity was attained mainly by investments in larger and higher-speed equipment to meet the ever increasing demand for steel.

---

[1] Adopted from the author's presentation before The Eighth World Productivity Congress held in Stockholm, Sweden, May 23-17, 1993.

In the second decade (1970s), the improvement was achieved by investments in the continuous process aiming at energy saving, cost reduction and manpower savings, simultaneously, in order to cope with the slowdown of the world economy due to the 1974 oil crisis.

In the last decade (1980s), and especially in the last five years, the improvement was done chiefly by concentrating on production with more productive equipment while discarding or replacing obsolescent, less productive, equipment. Though the rate of growth of productivity has somewhat declined in the last 20 years or so, we may say that it is still appreciably high taking into consideration the lifelong employment system which is unique to Japan.

As a factor in the high productivity mentioned above, let us consider the blast furnace; the symbol of the steel industry. What effect does the blast furnace size have on the productivity? As the blast furnace size increased, the productivity improved. So, why are there still many blast furnaces which are relatively small?

30 years ago, blast furnaces were operated by the experience and intuition of the operators; controls were based on information obtained from only four measuring instruments. By contrast, today's large blast furnaces are operated by many measuring and controlling systems.

As the blast furnace is increased in size, it becomes difficult to control the reactions which take place in the furnace to ensure the stable production of iron.

Just imagine the difference between a small airplane and a jumbo jet. The cockpit of the jumbo jet is equipped with numerous instruments and automatic control systems to enable only two or three pilots to fly the jet and some 500 passengers safely. The same is true of the large blast furnace. In order to ensure stable production of iron, one must control the chemical reactions between ore, coke, and gas, the transmission of heat, and the distribution of raw materials in the furnace based on information obtained from many measuring instruments. Today, blast furnaces in Japan are operated by artificial intelligent (AI) systems with only a few operators. In other words, we no longer need a blast furnace *maestro*.

To that end, it is indispensable to make the accurate observation of phenomena by measuring and analyzing instruments which were formerly perceived empirically and to build the system of theories for understanding the results of observations.

We have considered the blast furnace as one example, the same can apply to other steelmaking manufacturing processes.

I want to emphasize that the Japanese iron and steel industry—although based on technologies and facilities invented in the United States and in Europe—has continually tried to improve, scientifically, manufacturing system performance. By transforming the human operator's tacit knowledge, clinical engineering skills, and esoteric art [aesthetic awareness and capability] into an exoteric art, explicit knowledge and theoretical engineering skills, we have not only refined equipment function and design, but also make its performance more efficient. The result has been the steady improvement in productivity of the entire industry.

Improving Productivity: Increasing "Knowledge of Manufacturing Technology."

Taking the steel industry as an example, I have explained that the fusion of knowledge of making things (esoteric art, tacit knowledge, and clinical engineering skills) is the scientific and engineering knowledge that underlies the improvement in productivity of Japanese manufacturing industries. Now, let me go into some more detail on this point.

In any manufacturing industry, the process of improvement is innovation to improve productivity is simply the process of creating "new knowledge." In other words, it is not too much to say that the improvement of productivity, or competitiveness, of any enterprise depends upon the ability to create "new knowledge," as well as the amount of "existing knowledge."

What makes up the knowledge involved in making some products? What must be kept in mind here is that the "knowledge" which any manufacturing company ultimately requires to maintain competitiveness is its "knowledge of manufacturing technology."

The "knowledge of manufacturing technology" does not simply refer to the technology for making some products. It refers to technology for making the *right things*, of the *right quality*, that the *market wants*, at a *lower cost*, in the *required amount*, at the *proper time*, with more safety, more comfort, and more friendliness to the environment.

This requires extensive knowledge, including marketing, process design, production, maintenance, training and education, cost management, and quality control. The "knowledge of manufacturing technology" is the integrated whole or those elements.

From time to time, we meet with an attempt to develop a traditional art directly into production technology with the aim of putting it to industrial use. I believe the reason why many of these attempts fail is that the skills are not refined into a technology or the technology is not refined into a viable manufacturing technology.

There are also venture businesses armed with scientific knowledge jumping directly into the field of manufacturing technology. It seems to me that they too find difficulty reaching an industry in which productivity really matters.

After all, in order for any manufacturing company to secure competitive productivity and carry on prosperous business on a lasting basis, it is important not only to convert it s esoteric art, skills, and scientific knowledge into "knowledge of technology," but also to integrate them into "knowledge of manufacturing technology."

Technical Management's Role in Creating "New Knowledge"

Improving productivity and consequent competitiveness in any manufacturing industry involves:

1) Increasing knowledge of skills (esoteric art), knowledge of

science (knowledge of seeds), and knowledge of engineering (knowledge of means), and then,

2)   Converting those elements of "knowledge" into "knowledge of technology" and further integrating them into "knowledge of manufacturing technology."

In the management of a manufacturing business, pursuing and pushing forth the above two points is the essence of technical management.

Many Japanese enterprises carry on *jishu-kanri* activities in their workshops.

This activity, which assumes daily productive activity of creating new esoteric art (tacit knowledge), was started with the aim of encouraging every employee to create esoteric art (tacit knowledge) and making them feel the joy of making things—a joy shared only by humankind.

The esoteric art (tacit knowledge) created in the process of *jishu-kanri* activity has been refined into the "knowledge of technology" by the manufacturing engineers.   Here you will see a marked difference from the traditional mass-production system in which human beings were simply regarded as a replaceable machine part.

The number of esoteric arts which have been created by *jishu-kanri* activity at Nippon Steel in the past two decades is shown in this figure (Figure 8).   It may be said that the strength of Japanese manufacturing industries comes in part from their unique system in which all employees, down to every worker, are encouraged to create new pieces of knowledge and engineers and researchers convert this knowledge into technology.

Sophistication of Technical Management

I was deeply impressed by a lecture delivered by Professor Claude Levi-Strauss, the French cultural anthropologist, at a productivity congress ten years ago.

In his lecture, Professor Levi-Strauss made a far-sighted remark to the following effect. "Productivity should not, in itself, be treated as an economic problem. Productivity involves social and ethical elements, and the quantitative view of productivity as a measure of the efficiency of production of materialistic products is *morbid* considering the social, cultural, and political imbalances it causes."

With the 21st century not far ahead, it seems to me that manufacturing industries around the world are beginning to be faced with the problem Professor Levi-Strauss pointed out 10 years ago.

It is evident that the conventional way of increasing knowledge in the field of manufacturing is no longer enough. Namely, the familiar ways of making products from raw materials, putting them on the market, and optimizing the labor, material, and money invested no longer guarantee the existence of any manufacturing enterprise. Technology alone no longer allows the manufacturing industry to meet the individual's diversified views of *value*, which can hardly be quantified, such as that of the environment and those of humankind.

Today, a heated debate surrounds the effect on humankind and on nature of an unprecedented growth of materialistic society in the past decades. Consequently, increasing emphasis is now being placed on such concepts as "harmony with humankind," "harmony with nature," and "harmony with resources." These signs probably indicate a shift away from "goods-focused" to "human-focused" thinking.

It may be said that we now face the problem of how we should meet the needs of a new society and market which will come into being from the above new concepts, or how we should create the required "knowledge of technology," within the framework of a corporate strategy adapted to the new concepts.

The attached figure shows the two concepts of the manufacturing: one is the "product-out" concept of assembling raw resources such as man, material, machine and money, then converting the material into goods, and then selling them in the market. The other is the "market-in" concept; taking information from market and social surroundings into consideration. It is said that the "market-in" concept is important for the manufacturing industry. However, I must say that the concept already has became obsolete.

In order realize the challenges of the "humans-focused" thinking, strong leadership and a corporate strategy is required. In this sense, all elements of the enterprise—management and technology—must have a common language to communicate with each other. Unless both parties have smooth dialogue, they have to compromise with each other by exchanging tacit consent.

(Note: the attached figures indicate a more appropriate "cycle of management" system). By incorporating "perceived external conditions" into business strategies and creating the "knowledge of manufacturing technology" appropriate to the business strategy, can lead to an improvement in the overall productivity of manufacturing industries of the future.

I will explain this using a simple example. Today, recycling waste materials has become a very important social problem. When this problem is tackled as part of corporate activity, say, in the steel industry, it is almost impossible to solve it merely by developing a technology for recycling scrap iron. Actually, solving the problem requires the understanding of society regarding the social and economic values of scrap, as well as the motivation and strategy of the recycling company. Only when corporate strategy and technical development are combined, can the recycling of scrap take on practical meaning.

Needless to say, the base on which the manufacturing business stands is "technology." Moreover, the competitiveness of any business enterprise depends upon its ability to create "knowledge of technology."

Nevertheless, the "knowledge of technology," which is largely in demand, is the "knowledge of innovation." This new knowledge is to be adapted to new articulated social and market needs, rather than to the "knowledge of [technical] improvement," for the field of manufacturing. In other words, research and development for creating "knowledge of [technological] innovation" is increasing in corporate importance.

As shown in my final figure, "research and development" activity is centrally positioned as a consequence. What should be kept in mind here is that R&D always watches out for new external and internal information, and then converts its technological ideas to corporate strategy in understandable terms. At the sam time, corporate strategy transmits its requests to R&D in understandable terms for researchers.

Namely, the conversion between "knowledge" and "information" expressed in terms understandable to both corporate strategy people and to R&D people plays an important role. It is said that creating the "knowledge of innovation," while effecting the conversion properly is the key to technical management of the future. Thus, in implementing such technical management successfully is the key to improving the productivity of manufacturing industries.

Finally, I would like to point out two important matters which should be kept in mind when implementing technical management to improve productivity of manufacturing industries in the future.

First, establish the proper attitude to realize the new concept.
In addition to "hard knowledge" created by using the analytical methods of objective view of phenomena (based on the rationalism of Descartes), consider also the "soft knowledge" created by the *syncretic* approach, which subjectively grasps various phenomena and different world-views and values, then integrates them to achieve a new insights and deeper understanding.

Second, is the concept of time. The author Alvin Toffler refers to this notion in his book "War and Peace," the importance of paying more attention to time. Society at large, as well as science and technology, are changing at an accelerated rate. And many human cultures who hold different views of "time" are converging. Today's yardstick for productivity cannot be tomorrow's measure, because the yardstick itself will change completely.

In other words, without the ability to respond to external conditions quickly, and then change behavior quickly, a business enterprise will find it difficult to survive.

Azusa Tomiura

SLIDES IN ORDER OF PRESENTATION

Technical Management

Plays an Important Role in

Creating New "Manufacturing"

(1) Increasing knowledge of skills
knowledge of science
knowledge of engineering

(2) Converting those elements of "knowledge"
into "knowledge of technology"
and "knowledge of production technology"

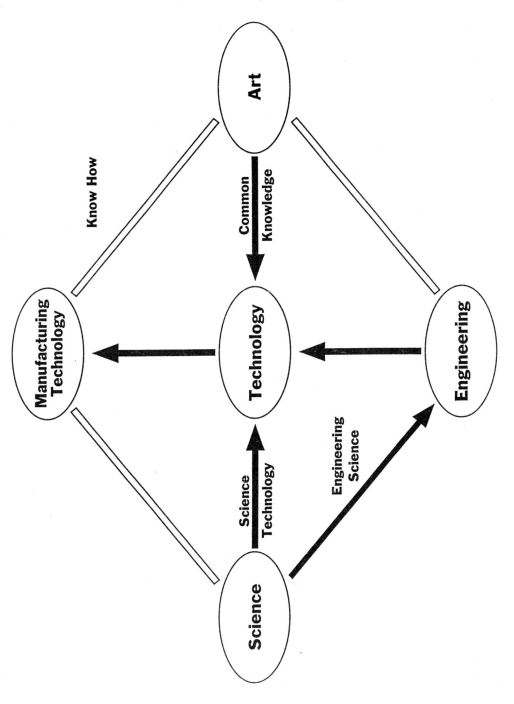

Structure of Manufacturing Technology

Marketing	• Market Assessment • Market Creation • Customer Service

Process Design	• Selection of Facilities • Process Layout • Selection of Process Route

Production Program	• Operational Practice • Reproducible Product Design • Improvement of Production Capacity

Maintenance Program	

Training and Education	

Cost Management	• Cost Assessment • Cost Improvement

Total Quality Control	• Product Liability • Quality Assurance

Harmonization	• Environment • Natural Resources • Human, Society and Safety

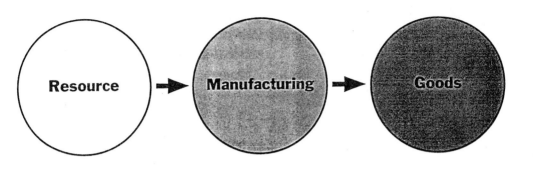

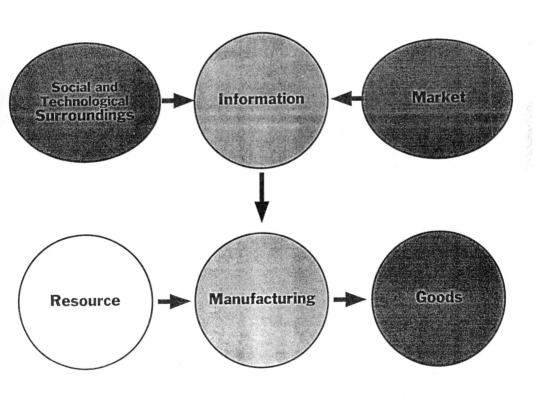

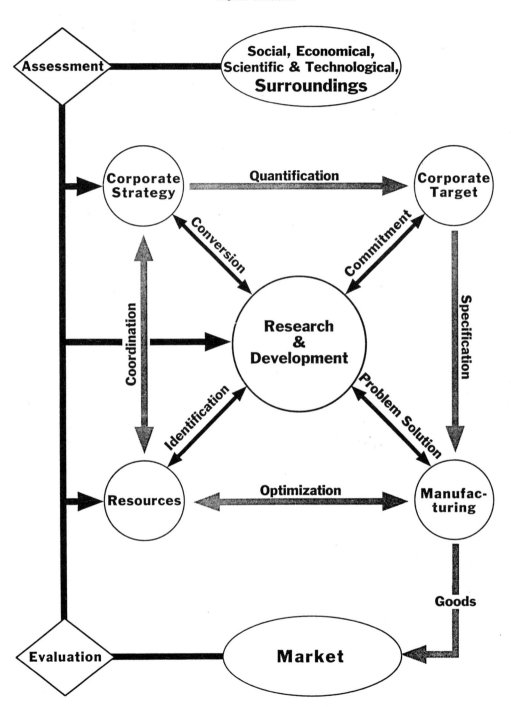

A MANAGEMENT VIEW OF JAPANESE CORPORATE R&D

R&D management is strategic human management, says an NEC executive who explains what that means in his company.

Michiyuki Uenohara

From an R&D management point of view, Japan is facing two major, simultaneous changes. One is that Japan has almost caught up with most European countries and the United States in industrial technology; hence, Japan has to break away from its traditional behavior as a follower and innovate new technologies and markets. The other is a transformation from an industrial society into a highly information-oriented society. In order to actively contribute to, and to get benefits from, such changes, R&D management has to change in accordance with new missions.

The Japanese industry, in high expectation of a prosperous information age, is hiring scientists and engineers (information technology engineers in particular), at a high rate. This is putting considerable strain on the engineer market. Although hardware productivity has improved considerably, software productivity is still low and the software industry finds itself in a dilemma of how to meet the market demand and improve productivity simultaneously. A similar dilemma exists among industrial R&D management.

In this article, I shall present my view of Japanese corporate R&D and discuss current problems of promoting basic research, meeting rapidly changing market demands, and managing R&D effectively.

R&D Investment Trends

Since the 1974 oil embargo, Japanese industry has made every effort to change from a heavy consumer of energy and mineral resources to a knowledge-intensive industry using less energy and resources. Almost all industries have aggressively invested in modern computer-controlled facilities in order to improve productivity and reduce energy consumption. Also, most non-electronics companies have extended their business spectrum into electronics.

Michiyuki Uenohara is executive advisor to the NEC Corporation, in Tokyo, and chairman of the Board of Trustees of the NEC Research Institute for Advanced Management Systems and of the NEC Research Institute, in Princeton, New Jersey. He joined NEC in 1967 after ten years with Bell Laboratories, and subsequently managed the Central Research Laboratories. He was elected to NEC's Board of Directors in 1976, and had responsibility for corporate R&D and engineering until June 1989. He is currently a Japanese member of the advisory panel established under the U.S.–Japan Science and Technology Agreement of 1988, a member of the Engineering Academy of Japan, a foreign associate of the U.S. National Academy of Engineering and a foreign member of the Royal Swedish Academy of Engineering Science. He holds a B.E. degree from Nihon University and an M.S. and Ph.D. from Ohio State University.

Electronic components are taking over functions from mechanical components, and software is taking over functions from hardware. This trend has heightened R&D competition in electronics, especially in information technology, making it far fiercer than it used to be.

This fierce competition is shortening product life cycles in every generation. Every competitor has to hire more engineers to develop, manufacture and market new products as quickly as possible. This is a kind of product development war. R&D management is demanding more engineers, but the supply is far short of the demand. The top corporate management demands an increase in R&D efficiency and, at the least, a halving of R&D time. The number of foreign engineers is increasing in Japanese industrial laboratories, but it is difficult to close the current gap between demand and supply. Also, a substantial improvement in Japanese R&D management is required in order to coordinate international R&D teams.

Measuring R&D Productivity

Many have tried to define and measure R&D productivity, but as yet there is no widely accepted definition and measure. R&D performance evaluation is mostly qualitative. Only at the final product development stage is the quantitative evaluation applied.

At corporate laboratories, numbers of technical papers presented at academic conferences and published in scientific journals, numbers of intellectual property right applications, and numbers of technology transfers and of consultations to business divisions are often used to evaluate performance. However, the tangible value produced is still unknown and its measure is qualitative in nature. It will take a long time until measurable tangible values are produced. Technological value cannot be determined within an organization. It is determined by the market and by competition. Therefore, the value is created not only by scientists and engineers but also by the management. I emphasize that the technological value is created mostly by management.

R&D outcomes are mostly knowledge and information. Consequently, R&D productivity depends largely on how broadly technology is applied. If a corporate division is tightly managed vertically from R&D to manufacturing and collaboration with other divisions is feared, R&D outcomes will be applied only to limited products within the division. However, if technology transfer is effectively managed across divisions, technology can be applied to many products in many divisions. The R&D

Reprinted with permission from *Research·Technology Management*, Nov-Dec. 1991, pp. 17-23. Copyright 1991, Industrial Research Institute, Inc.

productivity in the latter open division is much higher than that in the former isolated division. The open policy is especially important for corporate laboratories, where R&D is mostly oriented toward generic technology.

It is said that "Every failure is a stepping stone to success." Hence, every failure is a stepping stone to creating new technology. A revolutionary technology takes many years to establish. If the new technology is applied prematurely to products, it usually creates many problems in development and manufacturing. Such problems consume considerable manpower and time, and eventually lose product competitiveness. This kind of failure often results from mismanagement, pursuing a short-term R&D "efficiency" without understanding the nature of effective research and development.

R&D management is strategic human management. It is quite different from production management. R&D productivity, especially when it involves advanced research and development, should be measured over the long-term and over a broad spectrum. If productivity is measured in the short-term and for a narrow market application, the result will be very different and misleading. R&D productivity for generic technology should be measured in terms of effectiveness rather than efficiency. If the developed technology adds value to many products and has a strong impact on the advancement of science and technology, its productivity could be infinitely greater than R&D that is poorly managed for short-term objectives.

The quality of technology is heavily dependent on the quality and motivation of the researchers. However, the productivity and effectiveness of technology is largely dependent on the management, who can appreciate its value and help promote its applications for many products across organizational barriers. The management that pursues *efficient* R&D is usually lacking a cross-impact vision of generic technology and a broad knowledge of the market. Even the management of a basic research laboratory should have such vision and knowledge. Otherwise, R&D effectiveness and hence R&D productivity will be greatly reduced. R&D productivity is largely dependent on the R&D management.

Improving R&D Effectiveness at NEC

The key to competing successfully in industry is overall strength—each link in the chain from basic research to development to production to marketing must be as strong as the next. I believe that overall strength is the secret of the success of Japanese industry in general and of the NEC Corporation in particular.

As shown in Figure 1, the NEC Tree, our core business areas are communications equipment and systems, computers and industrial systems, electron devices and home electronics. These businesses contributed, respectively, 25, 44, 18 and 13 percent of our sales in 1989. As a technology-based company devoted to the integration of computers and communications, NEC is contributing to the betterment of the highly information-

Technological value is created mostly by management.

oriented society that is to come. We strongly believe that information, including software, will play an important role in this future society, but hardware innovation is basic to the improvement of information productivity.

NEC manages over 190 companies that form the NEC group. Employees total over 160,000, among whom about 35,000 are engineers. As the parent company, NEC alone employs over 38,000 people, about 15,000 of whom are engineers who handle administration, marketing and sales, R&D, and engineering.

In order to encourage these 35,000 engineers to communicate and collaborate more effectively, we have distributed NEC's R&D activity throughout the company and the affiliated companies. The Research and Development Group, the Computer and Communications Software Development Group, and the Production Engineering Group together make up the corporate R&D laboratories (see Figure 2). They belong to the operating organization rather than the staff organization. Ten profit-making operating groups have their own divisional laboratories and development departments. Many

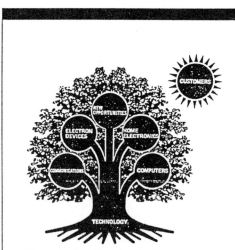

Figure 1.—NEC describes itself as a company that is rooted in technology and devoted to integrating computers and communications for the betterment of a highly information-oriented world.

affiliated companies have their own engineering and development departments.

The corporate laboratories are primarily engaged in long-term R&D for future products—not tomorrow's products, but the day after tomorrow's (see Figure 3). Every manufacturing division is responsible for developing marketable products—today's products—as well as the techniques necessary for high productivity and quality. The divisional laboratories work primarily on technologies that will create the group's near-term business—technologies for tomorrow's products. Of course, product and production technology are inseparable; even research for the day after tomorrow must proceed amid a strong awareness of production realities.

NEC spends more than 10 percent of sales, much more than the profit, on R&D in order to best serve customers and to secure company stability. Hence, about 90 percent of R&D expenditure is spent in the manufacturing divisions. Of course, it is difficult to define boundaries between technologies for today, tomorrow and the day after tomorrow. And even if they could be defined easily, the boundaries would be continuously changing as the market and the

> *Each link in the chain from basic research to development to production to marketing must be as strong as the next.*

technological environment change. To cope with such a complex situation, NEC has adopted a policy of selective diversification and concentration of engineering resources in order to facilitate close communication and cooperation (Figure 4).

The corporate laboratories receive about 65 percent of R&D funds from the corporate headquarters, about 33 percent from the manufacturing groups, and 2 percent from the government. The funding from the manufacturing groups helps to effectively transfer technologies and solve reliability and yield problems.

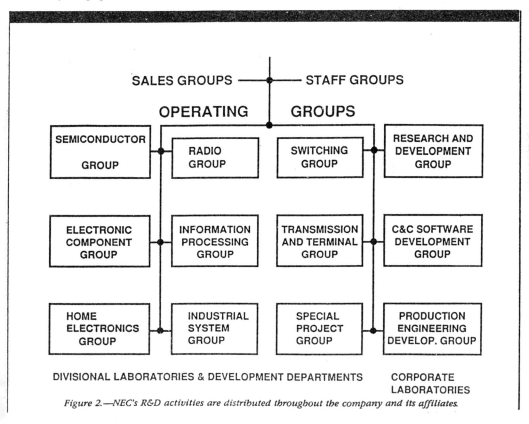

SALES GROUPS ——●—— STAFF GROUPS

OPERATING | GROUPS

SEMICONDUCTOR GROUP	RADIO GROUP	SWITCHING GROUP	RESEARCH AND DEVELOPMENT GROUP
ELECTRONIC COMPONENT GROUP	INFORMATION PROCESSING GROUP	TRANSMISSION AND TERMINAL GROUP	C&C SOFTWARE DEVELOPMENT GROUP
HOME ELECTRONICS GROUP	INDUSTRIAL SYSTEM GROUP	SPECIAL PROJECT GROUP	PRODUCTION ENGINEERING DEVELOP. GROUP

DIVISIONAL LABORATORIES & DEVELOPMENT DEPARTMENTS   CORPORATE LABORATORIES

Figure 2.—NEC's R&D activities are distributed throughout the company and its affiliates.

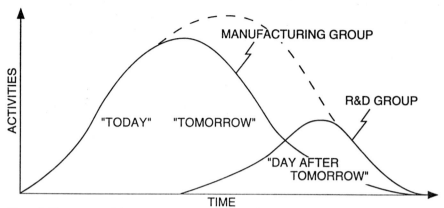

Figure 3.–The corporate laboratories are primarily engaged in R&D for the "day after tomorrow's products," while manufacturing is responsible for "today's products."

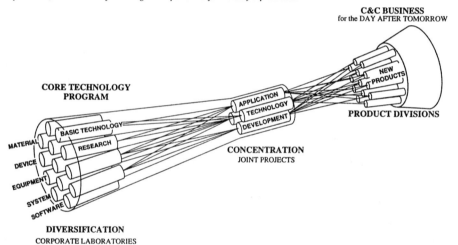

Figure 4.–To facilitate communication and cooperation, NEC has adopted a policy of selective diversification and concentration of engineering resources.

With such limited resources, how can the corporate laboratories meet their responsibility of creating new technologies for the day after tomorrow, transferring such technologies for tomorrow's products, and supporting customer services? We follow the diversification and concentration strategy shown in Figure 5.

Generic technologies are developed far ahead of

manufacturing divisions' needs in each professional laboratory. As soon as basic technologies are reasonably well developed, new core product models that effectively utilize such basic technologies are proposed to various manufacturing divisions. After hard negotiation and evaluation, joint development projects are initiated and development resources concentrated from both laboratories and manufacturing divisions. Laboratories provide key technologies and brains, and manufacturing divisions provide product development skills and established peripheral technologies.

Educating R&D managers to the point where they are able to identify the correlation between the market and the generic technologies, and then getting them to develop those technologies requires a time-consuming effort and long-range planning. On the other hand, the development of marketable products and the methods of producing them must be done in a responsive and timely manner. These activities consume a large share of engineering time and money. Indeed, if a thoughtful strategy is lacking, the education and basic research would be put aside and the commercialization activities would consume an even larger share of resources and time.

Formulating the strategy for R&D programs involves high-level interaction among marketing, operating and R&D groups. Considerable information must be gathered about future market trends, potential product ideas for business growth, and science and technology trends. This information is analyzed to determine future market potential—analyzed in the light of long-range corporate objectives. The heart of the strategy is the core technology program initiated by me 17 years ago.

Core Technology Program

The core technology program is developed from extensive analyses of both the market and the technology. A major analysis is done every 10 years for a two-year period, and minor modifications are made every year if necessary. The corporate laboratory top management is responsible for establishing the core technology program since it is the long-range R&D policy and important for establishing the corporate strategy from the technological point of view. However, since it requires extensive analysis and evaluation, the R&D planning office handles most of the process. The office members consist of a few managers, who are laboratory general manager candidates, a small number of analytical experts, and mostly senior researchers from every laboratory who hold two jobs concurrently—research leader and planner.

One-half of the researcher/planners are laboratory manager candidates and the remainder are specialist candidates. The manager candidates analyze market trends. Since the primary technology market for the corporate laboratories is every business division in the NEC group, the managers analyze the strength of all SBUs and relate their market growth potential to NEC's business portfolio. The specialist candidates analyze science and technology trends and list important technological areas. Based on these analyses, many correlation tables between key technologies and markets are developed. Also, new key technologies, which should be created or are missing, are itemized. These correlation tables identify key strategic technologies.

A major difference between our portfolio usage and common practice is that we define the core technologies

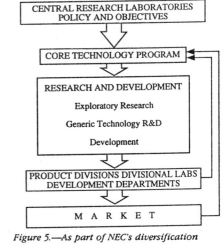

CENTRAL RESEARCH LABORATORIES
POLICY AND OBJECTIVES

CORE TECHNOLOGY PROGRAM

RESEARCH AND DEVELOPMENT

Exploratory Research

Generic Technology R&D

Development

PRODUCT DIVISIONS DIVISIONAL LABS
DEVELOPMENT DEPARTMENTS

M A R K E T

Figure 5.—As part of NEC's diversification and concentration strategy, generic technologies are developed far ahead of manufacturing division needs in each professional laboratory.

on which we have to concentrate strategically rather than discriminating among SBUs. Since the major responsibility of the corporate laboratories is to develop technology for the day after tomorrow, most corresponding SBUs should be in the "problem child" and "star" areas (see Figure 6). However, if the basic technologies developed are useful not only for such SBUs but also to reactivate the businesses of "cash cows" and "dogs," the benefits for the company are very great indeed.

During the course of core technology program development, the participating members acquire a broad spectrum of market and technological knowledge which most of them had paid no attention to before. They have also identified broad correlations between numerous products and basic technologies, and have created many new product ideas. Such knowledge is very useful to appreciate research activities by their subordinates and to bridge their outcomes with divisional product developments. The core technology development is useful for identifying strategic targets, but its educational benefits for manager candidates are far more important.

Strategic key technologies that should be pursued are grouped into a limited number of core technologies, each of which can be created and nurtured by a professional group. The strategic importance, minimum number of committed key basic technologies, important core products, and interdisciplinary relationships among other core technologies are defined for each core technology. The core technology program is presented to laboratory managers and divisional managers before the five-year plan is developed at each division.

In 1975, after two years of extensive study, we defined 27 core technologies. In 1983, since software engineering technology had been advancing so rapidly, we increased to 30 core technologies. In 1990, we increased to 34 core technologies. Many core technologies are nurtured jointly by several research laboratories. Some laboratories have to maintain close ties with many laboratories in order to establish their major core technologies. Core technology R&D management is by no means simple. The manager must have multi-dimensional management skills.

A group of core technologies demands a common strategy for better communication and collaboration. Hence, we have defined six strategic technology domains (STDs). The vertical management, or line management, is the responsibility of the laboratory manager; the horizontal management, or core technology management and STD management, is the responsibility of each specialist leader.

Effective Technology Transfer

NEC's product business is handled by ten manufacturing groups: four division groups in communications, two groups in computers, two in electron devices, one in home electronics, and one for special projects. These ten groups consist of about 60 product divisions and 25 divisional laboratories. Each product division has its own development departments for product technology and manufacturing technology.

Each product division is a profit center. Hence, they have to earn a profit and finance their own R&D. Therefore, they are very independent and also very competitive with one another. However, as electronics technology advances, and computers and communications merge inseparably, the basic technologies used in NEC's products become very common. If each division were to develop its own necessary technologies independently, many unnecessary overlaps in R&D would result, thereby wasting engineering resources. Not only would the resource be wasted but, due to the limited resources, the quality of technology developed would be inferior, and there would be product quality problems. This would result in a "patch-up" type of R&D and eat up future R&D resources.

To avoid such chaos, the divisional development laboratories help support and coordinate each division in the group; the corporate laboratories help support and coordinate, if necessary, all divisional laboratories and divisions. The concept and strategy based on the core technology program improves cross-organizational collaboration and cooperation. Those who participate in the development of core technology programs and those who are educated by them have been very helpful in effectively transferring technologies to manufacturing divisions.

Many manufacturing divisions and laboratories now cooperate to develop generic technologies by jointly developing core product models. These generic

Many manufacturing divisions and laboratories now cooperate to develop generic technologies by jointly developing core product models.

technologies and core products are commonly applied to develop practical products in harmony with marketing strategy. Product development lead time can be significantly reduced since most basic technologies and key components, which are necessary to develop any new product, are readily available whenever they are needed.

I believe the best strategy for meeting the diversified, rapidly changing modern market needs is to prepare appropriate basic technologies well ahead of the market development by corporate laboratories, and quickly respond to market demands by business divisions, integrating well-developed key product modules with new basic technologies. Here, the diversification and concentration strategy functions properly.

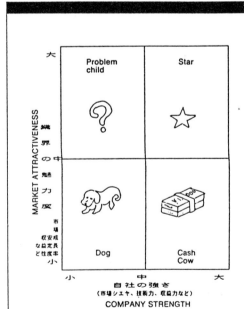

Figure 6.—Rather than discriminate among strategic business units, NEC defines the core technologies on which it has to concentrate.

Symbiotic Competition

I have often discussed the social roles of science and technology with our scientists and engineers. After several long debates, I defined my concept of true technological progress as being the popularization of frontier technology to make available appropriate hardware and software to anyone to reduce their mental, physical, financial, and social handicaps. New scientific discoveries, great inventions and technological breakthroughs are very important for technological progress. They are just beginnings, however, and not the goal of technological progress.

We have a common concern that increasing R&D expenditures will exceed the limits for every company and country. However, we cannot slow down technological progress. How can we cope with the dilemma? At the Third International Conference on the Management of Research and Development, in 1972, I presented a paper entitled "Symbiotic Competition—Future Trends of the Open Know-how Market." In this talk I stated: "As we recognize our own capability, we realize the need very clearly to develop our own specialties, while relying on others to support our weaknesses. However, a major question arises: How should broad research and development be covered? And what should be relied upon from the outside? In addition, we have pollution problems, and technological assessment is now loudly discussed.

"With such a range of problems, it is extremely difficult, if not impossible, to carry out the necessary research and development activity in large research organizations or

> *This is the time to promote global R&D productivity and to establish a rule for global symbiotic competition.*

countries. It may also be inefficient to carry out all research in one country.

"Symbiotic competition is, I believe, the future trend of the open know-how market. This is the only way we can maintain healthy growth and a peaceful society. What does symbiotic competition mean? It is mutual support and reliance for survival and growth within the basic rule of free competition."

I believe this statement remains valid today. We have almost established a basic rule for domestic symbiotic competition: to shake hands among friendly competing companies in the pre-competitive phase but stop as soon as one company starts to invest aggressively in market development. This is the time to promote global R&D productivity and to establish a rule for global symbiotic competition, in order that we may collaborate across national borders to utilize limited R&D resources effectively, develop excellent generic technologies, and meet market demands promptly. ⊕